Wild
SEX

Wild SEX

THE SCIENCE BEHIND MATING
in the ANIMAL KINGDOM

DR. CARIN BONDAR

PEGASUS BOOKS
NEW YORK LONDON

WILD SEX

Pegasus Books Ltd.
148 West 37th Street, 13th Floor
New York, NY 10018

Copyright © 2015 by Carin Bondar

First published by The Orion Publishing Group as *The Nature of Sex*, London, 2015

First Pegasus Books hardcover edition August 2016

ISBN: 978-1-68177-166-3

10 9 8 7 8 6 5 4 3 2 1

Printed in the United States of America
Distributed by W. W. Norton & Company, Inc.

To my Mom

TABLE OF CONTENTS

FOREWORD

What is sex? In terms of its biological definition, sex is the meeting of sex cells from males and females in order to create offspring. However, the process of sex is so much more than the mere meeting of sperm and egg. There are few aspects of life that remain unaffected by it – even for humans. Day-to-day functions like eating, sleeping and avoiding death are relevant only because they facilitate sex. In the biological world it means nothing to survive if you have not reproduced. But sex doesn't simply happen. You have to look long and hard to find the perfect partner. You have to make sure that you are as attractive to them as they are to you. Timing is critical, as your preferred partner may or may not be available for your sexual conquests at the exact moment you desire. What if your lover happens to want a kind of sex that doesn't interest you? Do you comply, or do you turn your attentions to other possible companions? What if they don't let you go? Sex and death are intimately linked and the desires of sexual colleagues are rarely aligned.

Is sex about companionship? Love? Passion? Procreation? The answer is all of the above and none of the above, depending on a plethora of biological and ecological factors. One thing is abundantly clear: sex is a critical part of the existence of any and all organisms on planet Earth. For humans, sex is comfort, fun, food, life. It can be both immensely pleasurable and immensely painful. We are lucky, because for us it's mainly pleasurable. For most other organisms it's predominantly the latter. As we explore the nitty-gritty of the diverse sex lives of animals, keep in mind the themes of biology and evolution. It may comfort you through what is frequently a pretty dark journey.

People often ask me, 'How did you become an animal sex biologist?' After all, it's not as though many small children daydream about becoming a leading expert on how animals get it on. The whole process was very organic for me; looking back it seems the fit was there long before I became aware of it. I defended my PhD while eight months pregnant with my second child, and she arrived soon after. At home with two little ones (which eventually became three, then four) I turned to writing and blogging to keep sane. I missed biology, I missed academic thought, and I was feeling tremendously isolated in our small Canadian town. I'm not sure what I would have done without the immense power of the Internet and social media, as my entire career has been borne from the ease with which people could find my work. I began by writing about a multitude of topics, always likening some form of human behaviour to the animal examples I talked about. However, one thing became abundantly clear: when sex was part of the story, people were always a little more interested (in some cases a lot more interested). I received a greater number of comments, questions and feedback when any aspect of sex, genitalia or mate selection was involved. The more gruesome the stories, the greater the reaction from the audience.

It seems people just love to hear about sex. We are all titillated on some level by the mere notion of sex, and chances are the majority of us spend a good portion of our free time either having sex or thinking about it. It's only natural for us to be curious about how the process takes place in other creatures. On the whole, human sex is straightforward and rather uneventful when compared to many other mammals and most definitely compared to the invertebrates. For us, the whole 'insert part A into slot B' is a fairly accurate description of what has to happen in order for successful (i.e. procreative) sex to take place. This *might* be the case for many other animals if it weren't for the notion of *biological fitness*. Scientists generally assume that all animals behave in such a manner so as to maximize their genetic representation in future generations (i.e. their biological fitness). From fruit flies to blue whales, organisms successfully navigate the costs and benefits of survival and reproduction in order to ensure that their own genetic blueprints are maintained in the population. This does not necessarily mean having as many offspring as their bodies will physically allow, because many aspects of ecology (i.e. predation pressure, environmental conditions, resource availability) vary through time and space. One of the most significant aspects of biological fitness when it comes to sex is that there is a huge discrepancy in the investments of females versus males.

Let's start at the beginning. Females have expensive sexual cells (gametes), the eggs. Think about it for a second with respect to humans. We ovulate once per month (with some given exceptions), and if the egg that we drop happens to become fertilized, we are out of reproductive commission for at least another nine months. However, our male counterparts produce an average of 180 million sperms *per ejaculate* (which could be viewed as a colossal waste, seeing as only one of them will be the successful fertilizer). In addition to the possibility of siring an offspring with the contents of a single ejaculate, most males are capable of ejaculating more than once in a 24-hour period, and are still ready to go again the next day. The bottom line is that sperm is abundant and cheap. Eggs are rare and expensive. These dichotomous standpoints between males and females often have repercussions on the level of sexual behaviour because it naturally means that females are choosier when it comes to selecting a mate. If a female has the ability to store sperm and has already received enough it makes much more sense for her to spend her time engaged in other biologically relevant activities, such as taking care of offspring, finding food or avoiding predators. This is, of course, bad news for any males who have yet to make a deposit into her sperm bank. Individual males would rather that females always accept their sperm donations, regardless of who has come before (pun intended), which sets the scene for some pretty drastic strategies for successful reproduction.

Although the 'love is a battlefield' theme is prevalent in the majority of organisms on the planet, we also see instances where sexual activities and the choice of partners are mutual decisions. When the level of parental care required is high from both sexes, males are expected to put a little more thought into where they are laying down their genetic blueprints (i.e. the quality of the female). In addition to the more conjunctive decisions made by some animals (such as some socially monogamous birds/mammals/primates) with respect to sexual partners, there are also cases where sex is actually enjoyable! Imagine that: recreational sex for the mere purpose of enjoyment and/or relaxation. Although it's relatively rare, it is certainly possible for organisms other than humans to engage in sex for indulgence and gratification.

The process of sex in the animal kingdom can be divided into three distinct categories. Before you even get a chance to mix gametes, you've got to find a partner. Perhaps humans take it for granted that we can quite easily come into contact with potential partners through our complex social practices. Most of us are blissfully unaware that for many organisms even coming into contact with a member of the opposite sex may be a challenge. The first part of this book will outline a small sample of the myriad ways in which merely finding a mate can be daunting and complicated.

Assuming you've successfully managed to find a partner, the next step is the act of copulation itself. There seem to be countless means of accomplishing successful fertilization of eggs. As I mention above, sometimes there is cooperation between parties, but all too often there is not. How do you successfully fertilize a partner who's not interested in being fertilized? The second section will outline the innumerable ways that this could take place, and we will also touch on examples where this conflict may be at its most intense: in organisms that are both male and female at the same time (hermaphrodites).

The third and final section of the book concerns the aftermath of what took place in the second one. After all, your attempts to maximize biological fitness could well be jeopardized if the offspring fail to thrive. Organisms may accomplish this through a massive parental investment into a low number of offspring, or they may conversely create a huge number of offspring in the anticipation that few of them will survive. Although certain kinds of animals may exhibit common strategies, there are often decisions at the level of the individual that can change the course of parental investment. For example, how should one behave if a better reproductive partner suddenly comes available? What happens to the offspring sired of poor-quality parents?

Putting together the three parts of this book (Meet, Sex and Aftermath) has given me a new appreciation for the enormous complexity that is involved each step of the way. The notion of normal when it comes to sex is completely impossible to define, which is an important point to keep in mind. There is really no regular way that sex can or should happen. The human-derived notion of what happens 'naturally' is about to get blown out of the water, because if we are going to label 'natural' as all the things that happen in nature, we've got to grasp that natural could mean stabbing one's partner in their forehead with a razor-sharp penis, or fertilizing juvenile females before their eggs are even mature. Horrifying? Yes. Natural? Yes. Although it may seem to be a near impossibility at times, the biodiversity on our planet should serve as a reminder that all animals are capable of successful reproduction. It may not be easy, but generally speaking, the deed gets done. The ways in which it gets done reverberate all the way through the various levels of individual, partner, family and population organization.

Sex has the power to influence every aspect of animal society because it requires males and females to behave in specific, characteristic ways towards each other. Communities where males predominantly coerce (rape) females have entirely different social stratification than societies where sex is either consensual or recreational. In societies where males cannot or do not coerce females, we may see a complete domination of the female sex in the social hierarchy. In addition to the act itself, sex plays a major indirect role in behaviours both between and within the genders. Strategic partnerships are created and reinforced by both the kind of sex and the specific individuals involved. Having a particular sexual partner can have effects that ripple up (or down) the social ladder. The same rings true in our own species: human sexual partners are brought into tribes or social classes that directly impact all aspects of future life – from the amount of money on the table to the context of indirect social interactions with non-sexual partners or the potential for resource acquisition and employment (or lack thereof). Sex is far from being an act that is in isolation from sociality, for humans or any other species. However, humans *are* exclusive in our disregard for the notion of biological fitness.

Our cognitive capabilities outside of the reproductive sphere have resulted in the near ubiquitous dismissal of the notion of passing on our genetic blueprints. *Homo sapiens* has developed a suite of ways in which *not* to create offspring so that we can free our lives to be busy with tasks other than those directly related to procreation while maintaining the ability to engage in sex. We actively seek partners that are willing to utilize some form of contraception to prevent the exact thing that every other animal on the planet is aiming to do. This is a fascinating example of where our species diverges from all others, although the extent to which we ignore our basic biology varies with culture and place. Despite the fact that we've taken the reproduction part out of the equation, the sex is still happening! Unfortunately, it's happening rather quickly, and in rather bland fashion when compared to the rest of the animal kingdom. To put it bluntly: human sex is downright boring. Male and female genitalia fit neatly into each other (for the most part) and the average time for males from insertion to ejaculation is generally less than ten minutes. Most human sex takes place in a horizontal position, on a bed or some other fairly comfortable structure, and is done in relative cooperation between partners. Compare this to a male that affixes his genitals to those of his female partner for days on end, requiring her to drag him around like some kind of deadweight sex toy. Or to females that are forced to comply with courtship and sexual intercourse while healing from predator-inflicted wounds or inhospitable environmental conditions? What about the male who has to rip off his own penis and spear it into a female's genital opening before she rips his head off? Folks, it's a tough world out there. Sit back and get ready for the titillating, exhilarating, horrifying, disgusting, alluring and wonderful world that is *The Nature of Sex*.

SECTION ONE

THE MEET

Generally speaking, it takes two to tango in the animal kingdom. But have you ever wondered how it is that we manage to pair up so effectively? How do we put ourselves into situations that facilitate meeting that 'special someone'? In truth, it's often more exhausting to meet an appropriate partner than it is to have sex with them. There are millions of single people in the human world: over 44 per cent of Americans over the age of eighteen are unattached. Not to mention those that were once attached and are no longer – divorce rates in the Western world are exceedingly high, up to 70 per cent in some countries. For humans, the decision of who to mate with is critically important, given that we generally have far fewer offspring than our bodies will physically allow, and we usually keep the same partner for all of them. So making the choice about the mate with whom we share our biological identity is massive. Do you take the plunge and pour your heart out in an online dating profile? Do you depend on the choices of friends or co-workers who set you up? The sheer number of ways in which human couples meet and pair off is huge, yet our options pale in comparison to the kinds of meet-up scenarios that occur in the rest of the animal kingdom.

After all, there is no wild sex on the agenda for one that has not found at least one willing partner . . .

Call me maybe

Human females are certainly influenced by the crooning abilities of our male counterparts. A sexy radio voice or a beautiful song can seduce us even without a visual signal to match it. Like many other sexual characteristics across a wide range of species, a male's vocalizations are a means by which to gain access to females. Many females rely on a male's song to gain information about his biological fitness, and they may be doing so without laying eyes on him. As with any kind of sexual ornament, a song has both easy and difficult aspects to its production and it is the difficult aspects that separate the men from the boys. In other words, anyone can tweet out 'Row-Row-Row Your Boat', but what about a stunning rendition of Beethoven's Fifth? Songbird females can judge a male on the length, dialect, repertoire and complexity of his songs. Even the consistency in the reproduction of similar notes, or a song's trill (where a single syllable is repeated in quick succession) is up for adjudication. Bachelor-birds generally sing different songs than those already paired with a female (you can insert your own joke here about the sad song of a 'saddled-down guy'). This makes a lot of sense because a male who has yet to find a mate has a completely different set of signalling priorities than one who is presently spoken for. *Presently* is the key word, because in most socially monogamous bird species, extra-pair copulations are the rule rather than the exception. This means that males of several bird species are constantly changing their tune depending on whether they are at home with the 'old lady' or whether they are out looking for a one-night stand.

In addition to creating mating calls alone, some bird species engage in a calling behaviour termed 'duetting'. This generally takes place in socially monogamous species, and is thought to function in demonstrating pair commitment. Duetting couples use distinct song codes to signal exclusively to each other, perhaps akin to the pet names or words that human couples share. It makes sense that such a couple-specific signal would serve to strengthen the pair bond in duetting birds, although there remains much debate about its biological significance, especially in light of the high incidence of extra-pair copulations that I mentioned above.

Birds are not the only organisms to utilize the sexual power of song. In fact the sheer complexity of songs utilized by organisms like amphibians and insects is astonishing. Male Emei music frogs (*Babina daunchina*) build small burrows at the edges of ponds where the female lays her eggs. The male tends to the fertilized eggs until they are tadpoles, and this again takes place within his burrow. Just how does a male attract a potential female partner? With song. Males produce courtship songs from both inside and outside of their burrows, although females have been shown to prefer calls that emanate from within. This is probably due to the fact that important information about the burrow can be gained from the acoustic properties of the songs. It's crucial for a female to evaluate both the male himself and the kinds of resources he's bringing to the table, including the area and depth of the burrow where she will make her ever-important genetic deposit. Male tree-hole frogs exhibit a similar level of acoustic prowess. They demonstrate the power of song by exploiting the properties of the tree-trunk cavities in which they nest. The cavities tend to be partially filled with water, which creates variation in the acoustics that result from a certain call frequency. Males will begin calling to potential female mates with a ranging pitch, but once they hit the frequency that results in maximum amplitude of their signal, they keep their calls steady. In this way, they maximize their chances of attracting a mate.

A little chemical help

A great number of organisms across the animal kingdom can be thankful for the chemical processes that bring them reproductive success. The power of olfaction (smell) is extremely important for a multitude of biological processes including predation, resource collection and of course finding a mate. Minute changes in the chemical signatures of various secretions can transform their ecological meanings. Although we *Homo sapiens* are not without natural signals of sexual chemistry (pheromones), we unfortunately spend a lot of time and effort clearing them away. Humans are under the incorrect impression that our natural scents are somehow dirty and unwelcome. We scrub our bodies with soaps and shampoos, and we use deodorants and perfumes and other unnatural chemicals to hide away what comes naturally. What people neglect to realize is that chemical cues are widespread, used by creatures from massive mammals to microscopic invertebrates, and for the most part they serve to bring receptive parties to the sexual arena. Many courtship pheromones have evolved for maximum efficiencies in both energy expenditure and gamete allocation. In other words, the chemical signals emitted by bachelorette #1 may be superior to those of bachelorette #2, resulting in a greater sperm allocation to the former. Imagine if you could identify an appropriate mate by simply smelling them.

The sheer diversity in the strategies of sexual chemistry is mind-blowing. There is a wide spectrum of approaches to sex via chemical help, from organisms who broadcast spawn (and therefore use chemical signals to ensure that their gametes are going to reach those of other members of their own species) to those that use a suite of chemical cues to maximize reproductive selfishness or to lure unsuspecting sexual victims.

We can generalize that male animals secrete chemical pheromones to attract females (and vice versa). This happens in many different scenarios and environments, both terrestrial and aquatic. A key issue for any particular male attempting to lure a female is that he'd prefer it if said female was only interested in *his* sperm. If a female is lured by the scent of a particular male, what's stopping her from being lured by the next, then the next, and so on? Evolution has dealt males a few distinct methods to deal with the problems that could arise in such a scenario. Not only do many males produce pheromonal cues to attract females (virgin females in many cases), they also produce chemical cues that cause a reduced or completely diminished response in the female once they have made their genetic deposit. Properties of oral secretions or ejaculates in diverse creatures show that anti-aphrodisiac pheromones often serve the purpose of redirecting a female's attention after mating. Essentially, such pheromones serve to enforce monogamy and to ensure paternity of resulting offspring. There are cases where this makes evolutionary sense for both males and females, such as when females receive all the sperm that they require from one sexual encounter and they can then turn their energetic attentions elsewhere. Males also produce and transfer pheromones to females that serve to decrease her attractiveness to other males. Again, this could be advantageous to females that aim to avoid harassment from unwelcome suitors, but this is certainly not always the case.

What happens when females are interested in additional copulations for increased genetic diversity in their offspring? What if she has been coerced into mating by a low-quality male? In these scenarios it does not make sense for a female to be put out of reproductive action by an anti-aphrodisiac or a pheromone that makes her 'ugly' to other males. Indeed, in many invertebrate species there is evidence for chemical 'arms races' where females have evolved resistance to male-generated anti-aphrodisiacs, which then evolve to compensate for these physiological changes and so on. Such biochemical warfare is complicated and extremely difficult to decipher.

There is more bad news for males utilizing the power of anti-aphrodisiac pheromones to increase their biological fitness. Several species of parasitic wasps, deadly parasites that infect larvae of many butterflies and moths, have evolved to take advantage of the chemical structure of these pheromones. For example, in a perfect world, a male cabbage white butterfly (*Pieris rapae*) transfers his ejaculate to a virgin female, and the anti-aphrodisiac chemicals contained therein put a stop to any further sexual behaviour on her part. She flies off in search of a suitable place to lay her eggs (oviposition) where she will deposit a clutch of approximately 20–50. This is where things go horribly wrong for both the male and the female butterfly. *Trichogramma* parasitic wasps have evolved a sensitivity for benzyl cyanide (the active ingredient in the butterfly's anti-aphrodisiac) so that when it is detected in a mated female, they hitchhike along with her to the oviposition site and deposit *their* larvae directly into her eggs. This means bon appétit for the parasite eggs, and nighty night to those of the cabbage butterfly. Since there are several species of parasitic wasp that can identify anti-aphrodisiacs in several species of butterfly, there should be strong selection on this trait in both parasites and hosts. In other words, since there is potential for colossal damage to the butterflies, there will be an enormous selective advantage for those with the genetic ability to withstand the wasp attacks. These 'protected' butterflies will go on to rear the most offspring and become abundant in the population such that most of the population becomes resistant to the wasps. There will then be some individual wasps with a more effective detection system to the 'resistant' butterflies, and these will be the most successful at rearing offspring – and so on.

The pheromonal cues utilized by males do not always have anti-aphrodisiac properties. Sometimes they are there to signal being attractive – or having something attractive. Male gift-giving spiders will provide females with a trinket (usually a prey item) wrapped up in silk. It's not just regular silk either, it's special silk anointed with sex pheromones reserved specifically for this purpose. The chemicals on the silk cause the female to accept the gift, and assume a courtship posture, hence initiating the copulation process. She grabs the gift, he grabs her. The neotropical spiders that make these pheromone-soaked gifts have the choice of making either 'sex silk' or web silk, as silk extracted from anesthetized males and artificially wrapped around gifts does not elicit an acceptance response from females.

Not all females are so unabashedly materialistic. In many animal species it is the dominance rank of a male that has everything to do with his reproductive success. Sometimes, chemical signals make a male seem more attractive to a female – especially if he's unable to win her heart in the 'old-fashioned way'. Female Australian field crickets (*Teleogryllus oceanicus*), for example, prefer dominant males to subordinates. So what's an inferior male going to do? He's going to alter his chemical scent to make himself seem more attractive. Unattractive males can increase the concentration of a number of cuticular compounds (cuticular hydrocarbons or CHCs) that are associated with increased mating success. It is a little like wearing extra-special cologne to make oneself appear more attractive – although it's a physiologically derived cologne, not some kind of impostor designer-perfume.

Male orchid bees (tribe Euglossini, subfamily Apinae) provide another fascinating example of male-created perfumes. They create their own signature scents by gathering substances from a wide variety of sources, including flowers, fungi, wet leaf litter, old logs, resins, rotting fruits or even faeces. They store the scent pot-pourri in special pouches on their hind legs, and it is hypothesized that females (who mate only once with one male) judge the males based on the result. These complicated mixtures are not comprised simply of the most abundant components in the environment: males are careful to include a wide variety of things that are rare or difficult to find. For these reasons, the bouquet of a male is believed to be a reliable indicator of his genetic quality. This is unfortunate news for males who are unable to produce scents as deliciously sweet as their direct competition, and researchers have observed males attacking each other – even removing hind legs of direct competitors – in order to steal their scents away.

Competition between males for female partners is extremely common and often involves direct deception. As with the bee example, one male's loss can be another's direct gain. Red-sided garter snakes (*Thamnophis sirtalis parietalis*) hibernate in large groups (tens of thousands) over cold Canadian winters. Upon waking, the snakes rapidly move into reproduction mode, and their massive squiggly orgies are quite a tourist draw for many small towns in the province of Manitoba. When the snakes first emerge from their chilly slumber, they are stiff and cold, and therefore more susceptible to avian predators like crows. Males will emit *female* pheromones to draw in some snuggle-time from other (already warmed) males and so speed their warming process. This kind of homosexual foolery has direct benefits to the males that receive the warming; they quickly stop emitting the pheromones once their body temperatures are appropriate for courting the ladies. Researchers have been able to induce pheromone emission from males by experimentally lowering their body temperatures, meaning that the trickery is confined to the early stages of emergence from hibernation. However, it's not all good news for the deceiving males. Courtship by several would-be suitors can negatively affect respiration rate, and in extreme cases males may be faced with forced copulation or death by suffocation.

I've done a lot of talking about the various chemical cues used by males to increase their reproductive success. What about those produced by females? In the majority of cases a female produces pheromonal cues as a means of communicating to prospective partners that she is sexually mature. For many invertebrates where prospective partners have to signal to each other from substantial distances, it's often the pheromonal cues produced by the female that are most instrumental. In addition, the scent markings of *virgin* females are markedly different from those that have already mated. This is important information for prospective male partners, who most often prefer a virgin female to one that's carrying the sperm of another suitor (or several). Perhaps the most drastic example of when a male should be picky about whether a female is a virgin is with sexual cannibalism. For males of many sexually cannibalistic species there is only one chance (or sometimes two) to be reproductively successful prior to getting munched. Therefore, it makes more biological sense to hold out for a female that has not yet mated. The primary way for a male to recognize such an important characteristic is through pheromonal cues. Indeed, for some orb-weaving spiders a male's exposure to female pheromones is the main factor involved in his willingness to be cannibalized. When exposed to mated females, males attempt to escape much more often than those exposed to virgins. The latter show a tendency to self-sacrifice in the name of biological fitness. This sharply tuned ability to detect changes in the pheromonal profiles of the opposite sex is a far cry from the human practice of washing them all away.

What's your sign?

When one thinks about the sexual displays used by animals (including humans) to impress each other, tangible things like physical appearance come to mind. For visually oriented species like ourselves what initially draws us to a potential partner is how they look on the outside. We might like to stress that other aspects such as personality, intelligence or sense of humour are also important, but the initial draw is usually physical attraction. This can be a tricky business for our species in particular, given that we utilize all sorts of unnatural beautification techniques to enhance our looks. For most other organisms physical appearance is an honest and reliable representation of overall health and genetic potential. However, it is becoming increasingly obvious that animals also rely on aspects of 'personality' when it comes to mate choice. A discerning female may not always be drawn to the brightest, shiniest, or most vocal male; she may choose a mate for his disposition rather than his physique.

A growing number of studies on the ecology of 'animal personalities' show that there are distinctive behavioural types within species, and these variations affect the amount of sexual action received – regardless of other aspects of physicality. In the past, such individual differences in behaviour have been dismissed as 'noise' around a general mean. As data accumulates, though, it has emerged that distinctive animal personalities form the basis of many aspects of behaviour, including reproduction. They are incorporated into populations because of distinct preferences (i.e. non-random mate choice) exhibited by both males and females within a species.

For example, male Siamese fighting fish (also referred to in pet stores as 'Bettas', after their species name: *Betta splendens*), adopt one of three distinctive behavioural types when they encounter a male and female together. They are lovers, fighters or dividers. Lovers direct all their attention to the female, fighters direct all theirs to the males, and dividers do a little of both. Females overwhelmingly choose to mate with lovers over fighters and dividers, probably due to the fact that over-aggressive males are liable to injure females during courtship. Why choose a mate who is going to take his overabundance of testosterone out on you? Similarly, female Japanese quail (*Coturnix japonica*) tend to choose less aggressive males – those that have lost fights with larger males prior to courtship. These 'loser' males may not have access to all of the resources that the 'bullies' do, but again, they are not as aggressive towards their female partners. In these birds, courtship tends to consist of males chasing a target female, continuously pecking her on the head and body, dragging her around by the feathers and repeatedly jumping on her back, so it comes as no surprise that females prefer the affections of a somewhat gentler suitor.

Mature male Coho salmon (*Oncorhynchus kisutch*) take personality differences to another level. They come in two distinct morphologies with two distinct personality types: jacks are small, and adopt a sneaky strategy when it comes to finding a mate. Hooknoses are large and aggressive, and tend to adopt a coercion sexual strategy towards females. Hooknoses have been observed to attack or even kill jacks if they come too close to a spawning female. They use their size and strength to mate with females – and often do so with great success. However, as we've seen with the examples above, it's not always brawn and machismo that win a female over. Female Coho show a distinct preference to mate with jacks over hooknoses, spending a greater time digging nests and in oviposition posture when jacks are the sole males present. (Both of these actions contribute to greater fertilization success for the jacks.) It's possible that females mate with hooknoses purely to avoid the high cost of coercion (biting, chasing, dragging), but they do what they can to increase the chances of fertilization to cooperative jacks.

Animal personalities are not simply a matter of aggressive versus non-aggressive types. There is substantial diversity in the kinds of behaviours that define a temperament, and attractiveness need not be associated with being tough or being weak. Female zebra finches (*Taeniopygia guttata*) appear to choose partners based on their level of exploratory behaviour, as do male great tits (*Parus major*). Both these bird species are bi-parental and socially monogamous, meaning there is careful selection by both males and females for potential partners. In this case it is not just about the personality of a potential mate but about the personality of the choosing party as well. Female finches and male great tits that are more exploratory themselves often choose mates from within their behavioural type – much as humans seek out like-minded individuals for adventurous activities like skydiving or river-rafting. A similar pattern has been observed in bridge spiders, where more aggressive males tend to mate with more aggressive females and vice versa. Bridge spiders are a non-sexually-cannibalistic species where males and females are of similar size.

So, do opposites ever attract? Indeed they do. When females and males both display distinctive behavioural types, like-minded individuals do not always make for perfect matches. In the tangle web spider (*Anelosimus studiosus*) aggressive males can easily beat out more docile males in male:male contests. However, such combative males are much more likely to be cannibalized by aggressive females, whereas docile males are likely to successfully reproduce with aggressive females, perhaps due to their ability to 'quietly' sneak in and copulate with minimum disturbance. On the other hand, aggressive males have a greater reproductive success than docile males when mating with docile females. This could be explained by the fact that aggressive males have no trouble out-competing their docile counterparts for females that are not threatening to them. In this way, such behavioural variation in populations is self-promoting.

Male orangutans (*Pongo* species) can be extremely coercive and violent towards sexually receptive females, so it makes a lot of sense for a female to do her homework and investigate the personality type of a particular male prior to mating. They do this by engaging in food-sharing with potential mates; or not so much 'food-sharing' as 'food-taking'. Females will attempt to get some food that is already in possession of a male, by grabbing it from his hands, feet or mouth. We're not talking about high-quality food items either; the kind of things she is apt to take are easy to find. In the human world the equivalent would be sneaking a few fries from a potential mate's plate. The female is not concerned about the items in question. In fact, for the most part she's not even hungry. She is taking the food merely to gauge the male's reaction – investigating aspects of his personality that indicate whether he will be an appropriate mate and/or father. The beauty is in the simplicity of the action – it's as if she's saying, 'I could easily get some of my own, but I'm going to help myself to yours instead. Deal with it.' In this way she's able to gain information that will help her select a mate. Males show a range of reactions to females engaging in this kind of behaviour, from simple tolerance to violent re-taking of the food, or even taking food from other nearby females. This kind of behavioural assessment goes a long way to helping a female select her ideal mate – although with orangutans (as well as with the Coho salmon and Japanese quail examples discussed above), coercive males often get a sizeable share of mating opportunities irrespective of female choice. The big tough guys are always going to be reproductively successful to a certain extent, but females of diverse species have evolved strategies to mitigate their dominance.

Plastic partners

As well as being required to make reliable assessments about sexual partners based on their physical and behavioural characteristics, animals (humans included) need to have the ability to revise these assessments based on any number of confounding environmental variables. Plasticity refers to both variability in ecological or biological factors, and variability in behavioural or physiological reactions that animals exhibit. For example, if anyone asks me about the most comfortable bed I've ever slept in, the answer is easy. After a long (*long*) day of hiking on a small island in Greece, I ended up in Corfu at around 1 a.m. – exhausted, dirty, and with nowhere to sleep. The hostels were closed for the night, and, being a twenty-year-old backpacker, I was on a very tight budget. Fortunately, my parents had provided me with a credit card to use in the event of a travel 'emergency', and when I weighed the options of either snoozing on a patch of grass in a park or finding a hotel, the latter won out. I can tell you without a shadow of a doubt that the bed in that hotel room was the most comfortable bed I've ever experienced. However, I'm fairly certain that if I were magically transported back to that bed today, I would realize that it was a middle-to-low-class mattress in a two-star hotel. This illustrates the extremely important concept of context dependence, and the same principles apply to plasticity in mate selection.

As if it weren't taxing enough for many males in the animal kingdom to produce the vocalizations, dances, physical structures or other mating signals for females to assess, environmental conditions have a bearing on their chances of success. On a warm summer evening at the beach after a picnic and a glass of Chardonnay we may be feeling carefree and indulgent towards members of the opposite sex. Conversely, if it's raining, our hair is messy and we're late picking up the kids from football, the exact same advances of the opposite sex may be unwelcome and annoying. Current environmental cues can be just as important as direct sexual signals when it comes to meeting a potential partner. Sometimes it is purely astonishing that any animals manage to get it right.

Male fiddler crabs (*Uca* species) have one enlarged claw that is used for a waving courtship display. Females prefer males with large claws and fast waves – or so it was thought, most research on the subject having been conducted under controlled conditions that isolated these traits. What happens when sex-selected characteristics are not isolated, and environmental variation comes into play? Fiddler crab vision is specialized to mudflat environments that are flat, clear and lack topographic complexity. However, some males make little mud-mounds upon which they stand to wave and advertise the entrance of their abode to females. This small change in elevation (less than 2 centimetres) has drastic negative effects on a male's mating success. Females show a marked aversion to males that signal from atop their tiny castles, despite both large claw size and fast waving frequencies. It can be assumed that a female's vision directs her pickiness towards males who are signalling on 'her level'. This kind of variation in female preference helps to maintain genetic diversity in male sexual signals because it means that the most physically capable males are not necessarily getting the girl. If waving from atop of a mud mound is so unattractive to females, we might ask the logical question – why do some males do it? It's possible that higher burrows are deeper and of superior quality such that if a male can lure a female into one she's more likely to stay. Higher males may also appear larger and be less susceptible to predators and may therefore represent a trade-off between survival and sex.

What happens when the environment is altered in such a way that sexual signals take on a whole new meaning? Phosphate pollution in near shore areas of the Baltic Sea has caused eutrophication – massive overgrowth of filamentous algae – in areas that were once clear with high visibility. These waters are prime breeding grounds for stickleback fish (*Gasterosteus aculeatus*); in a normal habitat the males build nests – the healthiest obtaining the most desirable territories – and wait for females to come and inspect them before deciding where to deposit their eggs. In a polluted environment, however, all bets are off. Eutrophication makes it difficult for females to find any nest, regardless of the male's territory or social status. In this thick algal overgrowth, weak and parasitized males can build their nests right alongside those of strong males and not be detected or chased away. Moreover, females tend to mate with the first male that they find. This is another (unfortunate) example where the strongest, fittest males will not necessarily have the greatest reproductive success. Whereas in the case of fiddler crabs such interference maintains genetic diversity, for sticklebacks eutrophication has the potential to direct 'natural' selection in entirely the wrong way.

Environmental variation can also wreak havoc on animals that use auditory signals to find mates. Ever tried to carry on a conversation with a potential mate in a noisy nightclub? Have you had much success in legitimate communication with deafening music and background chatter? I didn't think so. Imagine the struggles faced by animals that depend on communication via mating vocalizations against noisy backdrops. Even the most basic environmental perturbations can have profound effects on the ability of boys to get messages to girls. Indeed, this kind of phenomenon is aptly termed the 'Cocktail Party Problem' and has been an area of research in human sociology for many decades.

Several animal species have developed reliable mechanisms to highlight the sexual signal from environmental noise; however, not all sound pollution is created equally. Natural landscapes can be a veritable symphony, although animals have evolved in the context of the natural noises of the ecosystems around them. Anthropogenic (man-made) noise in urban landscapes is a much more recent phenomenon, and it is a major barrier to effective communication. Several species of urban birds change song frequency, volume, timing or duration of their mating songs so that they can reach females against a backdrop of traffic, industry and other kinds of urban buzz. Unfortunately, such adjustments are not always successful. In noisy urban habitats it is not uncommon for more males to remain unpaired during the breeding season – a simple matter of not being able to find an appropriate partner over the commotion of a city habitat. Indeed, there are few species that realize an overall *increase* in reproductive success in metropolitan areas. Those that are successful experience a completely different set of selective pressures than those in natural habitats, which can ultimately lead to the evolution of new species.

For example, urban-dwelling male great tits produce higher-pitched, faster-paced songs than those in rural areas, making the songs unrecognizable between populations. Similarly, grasshopper males in roadside (noisy) habitats produce different courtship songs to their counterparts in rural areas. They do this so as to be heard over the constant noise of the traffic, and the signals here are composed of a significantly higher frequency maximum than in quieter areas. Interestingly, the physiological mechanisms by which the grasshoppers have changed their tunes suggest that these new songs go a step beyond mere behavioural plasticity. We're not just talking about changes in tune, rather changes in the biological mechanisms by which sounds are produced. Grasshoppers from environments with different noise levels are on their way to becoming separate species.

It's far from easy to grab the attention of a potential female mate in a noisy nightclub, but things could be worse. Imagine trying to put your best courtship foot forward while riding a roller coaster. Environmental background noise can take many forms other than sound. Lizards that inhabit (and court females from) their positions on long grasses can find themselves exposed to strong winds that move the vegetation in unpredictable ways, making courtship an exercise in being sexy in conditions that are dangerous and downright nauseating. What's a head-bobbing, dewlap-inflating lizard bachelor supposed to do to woo a female against a backdrop of dramatic swooping grass? Jacky lizards in Australia (*Amphibolurus muricatus*) directly respond to changes in wind patterns by increasing the period over which signalling takes place. Anole lizards (*Anolis* species) increase the speed with which head bobbing occurs in order to enhance signal transmission on windblown vegetation.

If there's one thing we've learned from the preceding discussion, it's that the environment can do a lot of things to decrease your chances of sexual success. Even if you are a strong biological contender with an honest mating signal, the environment can be your Achilles' heel when it comes to scoring a mate. However, the environment can also do the exact opposite, if used wisely. In some cases you can increase the impact of your mating signals by grabbing a little help from Mother Nature.

For many bird species, having genetically programmed gorgeousness is not enough; males use the environment to put their best feather forward. Great bustard (*Otis tarda*) males augment illumination of their white plumage by pointing it at the sun, which maximizes the contrast of their achromatic (white) feathers against the dark background of their habitat. Iridescent plumage is a common component of sexual signals in many bird species, and such coloration is strongly influenced by ambient light. Basically, iridescent colours change in appearance based on orientation and exposure to the sun, so a male can make himself look much more attractive by posing in direct sunlight. Blue-black grassquit birds (*Volatinia jacarina*) in Brazil have been shown to instantaneously optimize their sexual signals by increasing display behaviour and vocalizing whilst bathing in direct sunlight. A similar use of the sun's rays has been demonstrated in male peafowl (peacocks) that display their plumage at a 45-degree angle to the sun during pre-copulatory mating rituals. The iridescent eyespots on the peacock's feathers are particularly showy when exposed to sunlight in this way, especially to females positioned directly in front of them.

Orienting to sunlight is not exclusive to avian species. Male jumping spiders (family Salticidae) display condition-dependent red faces and green legs to grab the attention of choosy females (who prefer the bright colours). If the male's red coloration is experimentally blocked, their reproductive success diminishes significantly. In addition, they use the power of the sunlight to enhance the appearance of their coloration and further increase reproductive success. Without this 'secret weapon' males are at a distinct reproductive disadvantage. Male guppies (*Poecilia* species) have been shown to increase their courtship activities in both the early morning and late afternoon, which coincides with light levels that would make them appear most attractive to females while minimizing detectability by predators.

Heat is another important aspect of the sun's power that can be harnessed to a male's advantage. In some lizard species, females decide where to live and breed based on information contained in scent markings of suitable bachelors. Access to thermal resources (i.e. a warm spot) is critical to reproductive success in these and other ectothermic (unable to regulate their own body heat) organisms, so a female needs to choose a male with a territory that allows for basking. Females have the ability to detect thermal-induced variation in the chemical composition of male scent markings, which enables them to make choices to maximize their reproductive success. In this way, males are indirectly signalling to females about the quality of their territories.

Predation is another common aspect of life for most members of the animal kingdom. I'm sure we can all agree that it's difficult to concentrate on wooing a partner when one's life is at stake; unfortunately, the showy sexual signals of many males have the side effect of making them even more noticeable and attractive to predators. This is a classic evolutionary trade-off between the demands of natural selection versus those of sexual selection. Consider the conundrum of both male and female swordtail fish (*Xiphophorus* species). The male's tail (sword) is a sexually selected characteristic. Females prefer males with long tails; however, in the presence of a predator, females associate with short-tailed males instead. This predator-induced plasticity in female choice results from the fact that males with large tails are more conspicuous, making them more susceptible to predation. Not surprisingly, females associating with said males are also at risk of being eaten, so when predators are around it makes more sense for them to choose a mate who's a less obvious target. Interestingly, this altered preference of females is short-lived. They generally shift their attention back to long-tailed males a few hours after the threat of predation has diminished.

The response of highly decorated males to predation threat can be plastic to a greater degree than the responses of their 'less showy' counterparts. Wolf spider males (family Lycosidae) can be either ornamented with large black brushes on their forelegs that are made even more enticing by a vibratory display, or without ornamentation. When faced with a predation threat, both showy and non-showy males cease their locomotory and courtship activities; however, brush-legged (ornamented) males take far longer to initiate courtship subsequent to a predation threat compared to their plainer counterparts. Perhaps not surprisingly, adorned males show a greater response to a predation threat because they have more to lose. If you've got expensive assets to protect, it makes sense to have a larger repertoire of responses to protect them. Except that sometimes the opposite seems to be true, as is the case with lesser wax moths.

As with any males in the animal kingdom, wax moths (family Pyralidae) vary in their level of desirability to females. The most alluring individuals have better body condition and attractive songs, which are both honest indicators of their biological fitness. Unfortunately for all male moths (desirable or not), their courtship songs are detectable by predatory bats. This means that an increased risk of predation comes hand in hand with attempts at courtship. According to a hypothesis called the 'asset protection principle', high-quality males should signal less often when conditions are risky because they have more to lose in terms of future mating opportunities, and this is what the brush-legged spiders exhibit. The same isn't true of moths: high-quality males begin signalling to females more quickly after a predation threat than their undesirable counterparts. Why? It's possible that in this case attractive males are exhibiting a 'live fast die young' strategy whereby they exhaust their resources and/or reproductive potential far sooner than less attractive ones. The beauty of biology is that there are exceptions to every rule.

Males of several fish species experience even more of a cramp to their dating style where predation threats are concerned. Adult male Trinidadian guppies (*Poecilia reticulata*) are brightly coloured, and their patterns of pigmentation are another classic example of a sexually selected trait. As we've seen, such conspicuous phenotypes represent a trade-off between the forces of natural and sexual selection – males that are brightly coloured and distinctive are clearly more susceptible to predation. If males are subjected to a predation threat during early development, their colour pattern emergence is both delayed and attenuated. Exposure to a predator during early development completely cramps a male's later mating opportunities, and masks what could have been an extravagant mating display. Male sticklebacks exhibit a similar attenuation of sexual signals in the presence of a predator. Their bright-red nuptial coloration is significantly reduced when the threat is high. What's the use of being gorgeous if you're not going to live long enough to reproduce?

SuperTramps

Biologists (myself included) harp on and on about the generalization that females are choosy and males are promiscuous. While this is a convenient rule of thumb, there are many scenarios where it simply doesn't hold. Consider polyandry – where a female mates with multiple males. It's not generally considered politically correct or ladylike in the human world, despite the fact there are many good reasons for females to welcome the affections of several gentlemen in close succession. Bear in mind that there are a few key elements to any sexual interaction. First, there are the behavioural elements, which entail selection and the act of copulation itself. Second, there are the genetic elements, which can involve any number of factors that allow for the successful fertilization of some sperms over others. Now, in species with nuptial gifts (see 'Trinkets and Tokens') it might make sense for a female to solicit a few extra copulations to reap the benefits of material goods that come along with them (behavioural). However, there are many species from a wide range of taxonomic groups where females actively solicit extra copulations from males without a nuptial gift in sight.

Biologists have long been perplexed by the existence of polyandry, because of the potential costs to females in terms of increased mating effort, exposure to sexually transmitted infections (STIs), and decreased time and energy for other biologically relevant tasks. So why do they do it? To begin, let's think about the act of sex itself. A major division between the hypotheses regarding the existence of female polyandry is to be observed in which sex is controlling the act of copulation. There are scenarios for when either females or males are the ones in control of the occurrence and duration of sex. Let's start with a discussion of the occurrence of polyandry when females are the ones in control.

Mouth-brooding female cichlids (family Cichlidae) are in control of both pre- and post-mating sexual selection. In this highly specialized mating system, females gather up their extruded eggs and brood them in their mouths while soliciting sperm donations from select males. It's like having several males ejaculate into your mouth in quick succession so that the most competitive sperm fertilizes the eggs. It's worth noting that this kind of mouth-brooding strategy is markedly different from most other fish that spawn into nests on the ocean floor. In the latter case, the female is *not* in control of which sperms are contributed for fertilization. Indeed, in many of these systems there are both dominant and sneaker males that have their own strategies for being reproductively successful irrespective of a female's desires. Subsequent to spawning, female mouth-brooding cichlids continue to solicit sperm donations, including from males that were not present when the eggs were first spawned. This sets the scene for intense sperm competition inside her mouth. She's effectively 'sperm shopping' in order to induce the highest level of competition; clutches can have up to four sires (eggs within a given clutch may be fertilized by different sperms). In addition to sperm shopping and mixing, females prefer specific phenotypic characteristics of males, such as those with elongated pelvic fins and large body length, so in this example it can be seen that females are in direct control of both pre- and post-copulatory sexual selection.

In many anuran (frog and toad) species, females have a sperm-storage organ (a spermatheca) that enables them to keep sperm for long periods of time. This too can allow for the possibility of female choice when it comes to post-copulatory selection of a particular seed. For example, female fire salamanders (*Salamandra salamandra*) can store sperm for several months, and they collect sperm deposits from many potential suitors during this period. In this scenario, there is a distinct decoupling between the processes of copulation, fertilization and larval deposition. In general, copulation occurs one *year* before a female gives birth to between thirty and fifty fully developed larvae, which she will deposit in streams and ponds during the spring. Females that collect donations from several males have been shown to have a much higher reproductive success than those that do not, indicating that multiple mating ensures a higher fertilization success. Similar trends have been shown for females of many species that are capable of storing sperm, including the Australian brown antechinus (*Antechinus stuartii*). In this small marsupial, females that seek the sperms of several male suitors have a three-fold increase in offspring survival over those that do not. It's clear that polyandry has the potential to increase the biological fitness of females through increased chances of successful fertilization, genetic compatibility and sperm competition.

These examples paint a pretty omnipotent picture of females. They are in control, maximizing their biological fitness by making use of the most eligible bachelors that the conditions afford. Female fallow deer (*Dama dama*) deliver only a single offspring per year and therefore have limited chances to get it right. They often seek the most dominant eligible bachelors for sperm deposits;

however, if too many females have 'come a-calling' he's liable to be sperm-depleted or may provide ejaculates with a more limited supply. With only one offspring per year it's vital for females to ensure successful fertilization, so they often engage in polyandry as a form of insurance; i.e. just in case the sperm from the first mated (and possibly more desirable) male doesn't make it into the reproductive tract, best to have a backup rather than lose an entire year of reproductive success.

Far from being choosy about potential suitors, green-veined white butterfly (*Pieris napi*) females exhibit a complete lack of selectivity, even when it comes to mating with their relatives. This failure to discern siblings from non-siblings prior to doing the deed means that incest is a regular occurrence. Both males and females show the same lack of inbreeding avoidance, which *could* result in substantial reproductive costs. Experimentally reared inbred larvae experience a 25 per cent lower hatching success and a 30 per cent lower survival rate to adulthood, which proves that mating with your siblings is a bad idea. Interestingly, females significantly decrease their re-mating interval when they've 'accidentally' copulated with their brothers, indicating that there are postcopulatory mechanisms in place to avoid the high costs of incest (although biologists have yet to determine the intricacies of what they are). So in this instance polyandry serves as a safeguard against having mated with a biologically inappropriate partner.

Polyandry has evolved in many species as a means by which females can obfuscate paternity of their offspring. In many organisms, and most specifically in mammals with a high degree of parental care, it's not unknown for aggressive males to commit infanticide on offspring that they have not sired. They do this for a number of reasons, the most common being to avoid providing parental care (energy/time) to an infant that is not biologically related, and to bring a maternal female back into oestrus so that he can impregnate her with his own seed. When the potential for such a horrific outcome exists, females actively solicit copulations from several males. Such behaviour has been observed in diverse organisms, including squirrels, primates and wolves, and is especially common in group-living species where males from neighbouring areas can immigrate into a female's social group and 'take over' as the alpha – killing all current offspring. Females attempt to mitigate these kinds of losses by keeping the males guessing as to who is the biological father of their progeny.

Although not as drastic as infanticide, males of many species harass or coerce females into mating. This is the unfortunate 'rule' in the animal kingdom, as opposed to the exception (see 'Sexual Coercion'). Many females have to put up with unwanted attention from their male counterparts, and many have evolved ways to deal with this. One such strategy is termed 'convenience polyandry', where females engage in (unwanted) copulations with would-be suitors for the simple reason that they do not want to spend the time or energy trying to avoid them. Convenience polyandry is prevalent where females cannot obtain direct benefits from multiple matings – genetic, material or otherwise.

Female green turtles (*Chelonia mydas*) are polyandrous, and multiple paternity of a female's clutch of eggs is commonplace. However, repeated mating in females is likely due to convenience polyandry because males are highly aggressive in their pursuit of receptive females. Multiple males often pursue a single female, and mating pairs can be harassed and/or bitten by males waiting impatiently for their turn. Additionally, unlike many of the previous examples, for female green turtles there is no increase in fertilization success, hatching or emergence success or increased hatchling size with multiple matings. Green turtles are capital breeders, meaning they largely rely on stored energy during their breeding period, so it doesn't make sense for a female to waste her precious energy fending off the amorous attentions of unwanted suitors. In this case the best strategy is to allow it to happen.

The story is vastly different in our smallest primate relative. Female grey mouse lemurs (*Microcebus murinus*) are extremely promiscuous during their short breeding season (just a few days out of the year). These ladies are of a similar morphological size to their male counterparts, and so they are not subject to intense intimidation or coercion from them. However, males are at their most abundant and eager during the female's narrow window of ovulation, and up to fourteen males can actively solicit a female at the same time. Curious as to whether female mouse lemurs were subject to convenience polyandry, biologists manipulated the diet of several females in an effort to create two experimental groups: those with high and those with low body conditions. They speculated that the females with low body condition would be more likely to engage in convenience polyandry due to a lack of energy to fend off the advances of interested males. Instead, the complete opposite appears to be true. High-quality females turned out to be the only ones to engage in polyandrous behaviour, suggesting a high energetic cost to such activities. This indicates that polyandry is an adaptive behaviour that is favoured over monogamy, and that mouse lemur females are in complete control of their sexual activities.

Sand tiger sharks (*Carcharias taurus*) have an extremely unusual mode of mating and embryonic development. Females exhibit polyandrous behaviour, but their babies exhibit genetic monogamy. In organisms with several eggs in a clutch (as we've seen with most of the examples above), having sperm from multiple donors means that eggs are usually fertilized by a mixture of the more competitive seeds. This is still the case for sand tiger sharks, but there is a distinct decoupling between the behavioural and genetic components of mating. The female ovulates over a period of several months, and fertilized eggs are collected in her paired uteri. These eggs (fertilized by a variety of suitors) all begin the process of developing into healthy babies; however, the first embryo in each uterus to hatch proceeds to kill and feed on the rest. This is known as embryonic cannibalism, and it results in a single massive hatchling emerging from each uterus. Thus, despite the female having indulged in polyandry and received genetic donations from many males, only one male's seed will be the winner in each uterus. So is this a case of convenience poly-andry – where the female appeases violent males so that the best sperm will prevail? Or is the female actively choosing the males she solicits so as to set the scene for a battle royale to take place in her uterii? As yet, biologists don't have conclusive evidence one way or the other. Shark mating is an extremely elusive activity, something we humans have yet to observe in detail for most species. We will have to wait for more robust conclusions on the rationale for female promiscuity in this species.

One of the biggest reasons why I love biology is that there are always examples that exist 'outside the box'. Biologists can devise rules and regulations for the costs and benefits of certain behaviours and trends, but every so often we see an example that defies them all. My last example of female polyandry is one such case. Male marine gastropods (snails) rarely provide any kind of post-sperm transfer investment in offspring. After copulation is complete, males tend to go their separate way while females lay their eggs on any number of benthic substrates (i.e. rocks or shells on the ocean floor). However, the snail *Solenosteira macrospira* is a major exception to this rule. In this species, males provide a substantial effort in terms of parental care. Females deposit their fertilized eggs directly on to the shells of males, transform-ing them into brooding grounds for developing offspring. Their shells become completely covered with embryos, which carries a significant energetic cost to the dutiful dads. But here's the kicker: the male is genetically responsible for something in the region of 25 per cent of the embryos he is carrying. Females of this species exhibit one of the highest levels of polyandry seen in an internally fertilizing organism, utilizing the sperm of several male donors to fertilize their eggs. On average, the load of babies being carried around by a single male will have been sired by between six and fifteen males. So in this instance females are making use of the genetic benefits of multiple matings to the direct *detriment* of the males carrying the embryonic load. How could evolution 'allow' for such an apparently unfair scenario to unfold? That's a question biologists are not yet in a position to answer. Which is why I love this example. Sometimes, nature doesn't make sense to us, and that's the beauty of it.

Monogamy. Really?

Monogamy has long been a topic of sexual behaviour that I find both fascinating and frustrating. Why are humans largely monogamous? Is it dictated by elements of society, rather than biology? Divorce and infidelity rates are high enough for me to seriously question the notion of having only one sexual partner for any kind of substantial timeframe, yet our huge brains and emotional nature make juggling many partners a near impossibility. So when we look to the animal kingdom, what do we see? Not surprisingly, we see a wide range of social and sexual relationships, some involving monogamy. For many decades it was believed that social and sexual monogamy were one and the same. Reports of wholesome 'family' living and sustained relationships with single partners were described for many species of birds, fish, amphibians and reptiles. However, a major shift in our understanding of monogamy took place with the advent of genetic sequencing. Once it was possible for us to examine whose babies belonged to whom, it became blatantly obvious that social monogamy and sexual monogamy were two separate things.

Without question, many animals associate with the same partner over a breeding season or over consecutive seasons. Partners

35

engage in both social and sexual activities with each other; but the majority of partners are also involved in sexual activities outside of the pair bond. And why not? When it comes to sexual reproduction, maximizing one's biological fitness is the ultimate goal, and reproducing with only one partner is not the most efficient way to achieve that (in most cases anyway). 'Monogamous' relationships run the entire gamut from complete, total social and sexual fidelity (rare) to massive levels of cuckoldry and sexual sneaking around. I'm going to start the discussion with the former and proceed towards the latter using examples from the animal group where monogamy is most often observed – birds.

The life history strategies of many bird species lend themselves to pair bonding because there is generally a good deal of parental care involved, requiring the efforts of both mum and dad. Here we see deviations from the usual sexual strategies of females being ultra picky and males being ultra enthusiastic, because if *both* partners are going to be sticking around to help with the offspring, both will tend to be choosy. Indeed, strategies of mate selection in birds often involve examination of both males by females and females by males, and both males and females exhibit secondary sexual characteristics. But first let me reiterate that this is not true for *all* bird species, as we also see many examples of extreme male showiness in response to extreme female choosiness in species where males do not offer any kind of parental care (think peacock). However, in a large number of bird species, monogamy in some form applies.

At one end of the spectrum are examples of total monogamy, such as that observed in scrub jays (*Aphelocoma coerulecens*) in Florida. When we ascribe monogamy, we tend to approach species on a 'blanket' level, i.e. if a species is sexually monogamous, then they are always 100 per cent sexually monogamous. This is a rather unrealistic way to approach any facet of animal behaviour, because there will usually be exceptions to the rule. As I've discussed in earlier parts of this book, ecological context plays a key role in the potential courtship and reproductive capacities of many organisms, and we would therefore expect a similar variation when it comes to sexually monogamous tendencies. This is why biologists interested in the level of sexual monogamy exhibited in scrub jays examined it in a range of contrasting habitats including, 1) a fragmented site where suitable scrub jay habitats (with suitable prey abundance) of various size are embedded in a random matrix of inhospitable ones; 2) an optimal, continuous and natural habitat with high prey abundance; and 3) a suburban habitat with high human influence and low levels of arthropod prey. Such differences in environmental stress, food availability and perceived parental investment are hypothesized to impact whether partners engage in sexual activities with others. For example, in more inhospitable areas like the suburban environment where prey availability is low, we would expect parental investment to be higher for both males and females, reinforcing the need for both social and sexual monogamy. Males are required to engage in activities like collecting food or finding suitable nest materials, and so they don't spend time or energy copulating with other females. If they were too busy copulating to provide food for their offspring the entire brood would probably starve. In the same way, a female in a stressful environment should avoid extra-pair copulations that would cause a male to question his role as genetic father and abandon her, because without his provisioning the offspring won't survive.

The opposite is true in habitats with abundant cover and food. Here, the costs of rearing offspring may be lower and such dedicated parentage by both mum and dad is not essential to guarantee their survival. In this kind of scenario we'd predict that both females and males might be interested in spreading their genetic seeds a little further. Thus the scene is set for a range of extramarital sexual action in scrub-jay land . . . So what did the researchers find? DNA analysis of thousands of offspring from couples at all three sites revealed a complete, total dominance of genetic monogamy. Despite the environmental heterogeneity, scrub jays maintained both social and sexual monogamy. This is surprising and extremely rare. I'm not sure if we will ever completely understand the reasons behind such an extreme case, but biologists have suggested that the high level of competition for breeding vacancies could explain sexual monogamy in Florida scrub jays. Intense competition between males leads to a general high quality level of all males in a population, such that females won't encounter males of vastly different condition. If all males are relatively equal in terms of their quality as mates and fathers, it could lead to female fidelity because they cannot really do 'better' by seeking out another mate.

High rates of both social and sexual monogamy have been observed in Steller's jays (*Cyanocitta stelleri*) as well. In fact, long-term pairs tend to have a higher breeding success than newly formed couples. The 'mate familiarity effect' describes this phenomenon (observed in several bird species), where an increased biological fitness via healthy offspring production occurs in long-term couples. This could be due to a few different factors including that familiar partners are able to save energy through increased coordination of resource-gathering and parental routines. Additionally, staying with the same partner for repeated breeding seasons tends to mean staying on the same territory, so there may also be fitness benefits to be gained through familiarity with one's habitat. This raises the important question: is she staying for me, or for my territory? It's a legitimate query, seeing as in many polyandrous species males are judged extensively on their ability to out-compete other males for high-quality territories on which to breed and house their mates. Researchers investigating the effect of territory quality on the extent of mate fidelity in various species have found contrasting results. Female long-tailed manakins (*Chiroxiphia linearis*) and Uganda kob (*Kobus kob thomasi*) (both bird species) tend to remain faithful to their mating *sites* rather than the male partners within them. On the other hand, a female satin bowerbird (*Ptilonorhynchus violaceus*) will not mate if a new male is present at her familiar bower site, while the female great snipe (*Gallinago media*) will seek out previous male partners even if she travels to different territories in subsequent breeding seasons.

Both of the jay examples described above are fairly uncommon in the extent of their social and sexual monogamy. Generally speaking, in socially monogamous birds we almost always see a low-to-moderate level of extra-pair paternity, indicating that the female is getting some sexual action on the side from cuckolding males, while the male is potentially out doing some cuckolding of his own. There are examples of bird species where the level of available habitat or resources *does* affect the level of sexual monogamy, and there's a compelling example from the mammalian world as well. Arctic foxes (*Vulpes lagopus*) are socially monogamous, and males have a high investment in care of offspring. In conditions of abundant food availability, high levels of extra-pair paternity exist (up to 31 per cent). When resources are less abundant, and hence when females may require more help from their male partners in the harsh Arctic climate, they are less likely to engage in sex with other male partners. In addition to being influenced by resource availability, the extent of sexual monogamy in socially monogamous couples may also depend on the current adult sex ratio. Perhaps it's inherently obvious, but localized variation in the number of possible partners can play a part in whether a male or female seeks extra-pair copulations.

The 'divorce' rates of birds in socially monogamous partnerships are widely variable, from 0 per cent in swifts (family Apodidae) and wandering albatrosses (*Diomedea exulans*) to 100 per cent in house martins (*Delichon urbicum*) and grey herons (*Ardea cinerea*), and are often dependent on operational sex ratios. When females are more abundant, males are more likely to divorce their partners in search of another. The opposite is true for a male-skewed sex ratio, where females tend to seek extra-pair copulations. However, when males are more abundant, and therefore not all males in a population are guaranteed sexual action, the incidence of forced copulations (i.e. bird rapes) can also increase and cannot be discerned via genetic analysis alone from females seeking extra copulations. Indeed, a study on wandering albatrosses found that rates of extra-pair paternity were between 14–24 per cent during three seasons of observations on Marion Island (Antarctica). This was due to a combination of both females seeking out additional copulations, and sexually frustrated males forcing themselves upon them.

Among the bird species with the highest levels of sexual sneaking are blue tits (*Cyanistes caeruleus*) and tree swallows (*Tachycineta bicolor*), where up to 90 per cent of broods are commonly found to contain offspring sired by males other than the female's social partner. Females actively solicit copulations outside of their social pairs, and maintain a commanding level of control over genetic paternity of their offspring. During times of environmental stability female tree swallows are less likely to seek extra-pair copulations; when environmental conditions are unpredictable or poor, though, they make sure that their broods are genetically diverse by seeking the genes of several suitors. Determining conditions are not confined to immediate aspects such as weather or prey availability: other factors like whether it's early or late in the breeding season, group-breeding synchronicity, parasite load or level of anthropogenic disturbance can all impact the level of extra-pair paternity found within a brood.

The bottom line with respect to almost all 'fathers' in socially monogamous bird species with high levels of sexual polygamy is this: males provide care to offspring they have not sired. This seems, on the surface anyway, a poor choice from the point of view of the male's biological fitness, but there are several factors that need to be considered before drawing conclusions. First, it's possible that males simply cannot tell when they have been cuckolded and are raising offspring that belong to other males. Second, though a male may 'suspect' that some of the offspring in his nest have been sired by another male, since he himself may have been out engaging in some extra-pair copulations – leaving another male to provide parental care to those offspring – perhaps it all evens out. Third, pair bonds may not be easy or quick to form. In the case with the albatrosses mentioned above, the pair-bond formation period is extensive, meaning that it may make more biological sense for a couple to remain together in the long term as opposed to either one starting afresh with a new and unfamiliar partner. So yes, there are scenarios where males will (perhaps knowingly) put their own resources into the biological well-being of other males by providing parental support to offspring that are not their own.

This is no big deal when compared to the virtually inexplicable 'parental' efforts of subordinate adult white-browed sparrow-weavers living in monogamous mating groups. Sparrow-weavers (Plocepasser mahali) are a group-living bird species that inhabits arid and semi-arid regions throughout Sub-Saharan Africa. They form territorial aggregations that engage in co-operative breeding, meaning that adults other than the genetic parents of the offspring help to raise them. There is general collaboration among the adults, with group members also helping in nest maintenance and territory defence. Groups comprise up to thirteen members, with relatively equal numbers of males and females. Now, here's where it gets interesting: there is only one breeding pair per group, a socially and (mostly) sexually monogamous male and female. All the other adults are subordinate helpers, and they do not attempt to mate with each other or the dominant couple. Subordinates play a vital role in the successful upbringing of the offspring of the dominant couple.

In some animals with a cooperative breeding system, this kind of reproductive forfeiture can be explained by the fact that the subordinates are in fact older offspring of the power-couple. By helping to raise their siblings, individuals still maintain a high level of inclusive fitness. Indeed, the 'monogamy hypothesis' suggests that cooperative breeding can evolve when levels of promiscuity are low. This is because low levels of promiscuity will lead to increased levels of genetic relatedness between group members, making them more likely to help out instead of reproducing on their own – exactly what we see in the former scenario. Biologist E. O. Wilson has been quoted as saying 'Sex is an antisocial force in evolution' because an increased level of promiscuity (leading to decreased genetic relatedness of individuals) ultimately leads to decreased cooperation between group members.

Interestingly, the monogamy hypothesis does *not* explain the instances of subordinate helping in sparrow-weaver groups, where the helpers are not genetically related to the reproductive couple. In these aggregations the dominant couple maintains a complete monopoly over all breeding activities, and adult helpers contribute to the well-being of offspring that are no direct relation. Subordinate females remain reproductively quiescent despite the fact that they are a similar size and age to the dominant female, and they have a similar body condition. Why on earth would they do this? Why not sneak into bed with the unrelated and sexually mature male who is right beside you, not to mention helping you in caring for the offspring of two other completely unrelated individuals? It's possible that the subordinate males and females forego attempts at reproduction for fear of backlash from the dominant couple, and the threats to their potential offsprings' fitness that may take place if they went ahead with it. After all, there are clearly defined social roles here, so it follows that the dominant male/female could threaten alien offspring, or that the offspring of the dominant couple would be superior competitors to those of the subordinate couple. It's a scenario that has yet to receive a logical explanation in biology – because *any* reproductive attempt by an unrelated couple would (biologically speaking) be more advantageous than devoting all of one's efforts to taking care of unrelated offspring.

The discussion here has largely focused on bird species because this is where the prevalence of vertebrate monogamy exists. Mammalian species exhibit a drastically reduced incidence of social monogamy: it's seen approximately 5 per cent of the time, and sexual monogamy has only been documented for five species (owl monkeys in the genus *Aotus*; the California mouse *Peromyscus californicus*; Kirk's dik-dik *Madoqua kirkii*; the Malagasy jumping rat *Hypogeomys antimena*; and pack-living coyotes *Canis latrans*). That's not to say there aren't more potential cases of sexual monogamy among mammals, merely that they have yet to be confirmed though genetic data.

There are a number of social and physiological reasons for this lack of monogamy. Most critically, mammals have internal gestation of offspring by mothers, followed by a period of lactation whereby infants are dependent on mothers for all their nutritional needs. In addition, primate infants are largely altricial – meaning that newborns are dependent on their immediate caregiver (mum) for most aspects of survival. Our very large brains have a lot of developing to do outside of the mother's womb (if they were any bigger prior to birth they wouldn't fit through the birth canal), which means that newborn primates are pretty pathetic. As a result, there isn't much that a mammalian father can contribute to his offspring's well-being from conception until weaning – which is drastically different from the situation faced by male birds, who can directly contribute to the gestation of their offspring by tending to eggs in the nest. Given that their direct investment during this phase is highly limited, male mammals are biologically better off siring offspring elsewhere . . . and as a rule, that's what happens.

That said, it's important to stress the difference between sexual and social monogamy, since there are a few examples of the latter in the primate world. Biologists have proposed a number of reasons for its evolution, including the 'female sociality' hypothesis, where range overlap in females is reduced due to feeding competition. This essentially means that females spread themselves out to different foraging territories, making it difficult for males to retain exclusive access to more than one of them. Males therefore have difficulty maintaining control over the sexual exploits of more than one female for the simple reason they cannot be in two places at once. Their only option is to choose to guard one particular female, making their sexual exploits monogamous (at least in the short term).

Another school of thought about the evolution of social monogamy in primates derives from the incidence of infanticide by males. Infanticide can be fairly common in some primate groups; approximately 34 per cent of gorilla infant deaths and 64 per cent of langur infant deaths are due to infanticide. In an effort to prevent this, females are thought to have evolved a socially monogamous strategy in several primate groups. Interestingly (yet not surprisingly, based on the wonders of biodiversity), other primate females have adapted the complete opposite strategy to deal with the potential of infanticide: paternity confusion (see 'Sexual Coercion' and 'SuperTramps' for further discussion on this topic).

Trinkets and tokens

Gifts are great. Who doesn't like to receive a special treat that's been specifically selected just for us? There are many humans that take gift-giving to a ridiculous level, pandering to our sense of selfishness and obsession with material goods. It's not taken to the same extremes in the animal kingdom, but the notion of gift-giving is commonplace there too. The differences in reproductive values of sperm versus eggs, and the resulting conflicts of interest that can ensue between males-who-want-more and females-who-have-had-enough, have resulted in males evolving a variety of ways to either change a female's mind or force her into sexual relations. In the former category, it's in a male's best interest if he can convince a female to have sex with him without resorting to violence. Even if a female has yet to meet her quota of sperm deposits, with many potential suitors waiting for their chance, it makes sense for a male to sweeten the pot a little. Good genes are not always enough. In many cases males have to prove their genetic superiority by producing a gift of similar quality. Then, and only then, do their sperm have a chance at obtaining the ultimate prize.

When we refer to this behaviour in invertebrates, we often use the term 'nuptial' gift (as though there is some kind of sweet, over-the-threshold wedding nest waiting for our happy insect couples). In reality, nuptial gifts can take many diverse forms, from prey items, to glandular secretions, water, specialized meals, even parts of a male's own body! The general idea behind the existence of these gifts is that they give the female something to do (i.e. *eat*) while the male engages in copulation and sperm transfer. The longer she spends eating, the longer his sperm has to reach her reproductive tract and the greater his chances of fathering the next generation. For many decades, biologists were under the impression that nuptial gift-giving was a mutually beneficial strategy in reproduction. Males win because they get a higher fertilization success, females win because they obtain some nutritious and tasty vittles that contribute to health and fecundity. Indeed, there are many examples where this kind of win-win strategy appears to be the case. When the gift given to a female is a prey item, she receives both direct (nutritional) and indirect (no foraging costs) benefits.

Females of several insect species like butterflies, beetles, crickets and dobsonflies realize an increased rate of survival when provided with such gifts. For some species of hanging flies and dance flies, ladies rarely go off in search of insect prey, so to obtain some from a male is a rare and welcome nutritional treat. Male nursery web spiders (family Pisauridae) have been observed to bring large prey-item gifts to females that have a positive correlation between maximum diameter and weight, indicating that males aren't puffing up a cheap gift, they are legitimately giving females an expensive present. The main point about all these gifts is that they are honest – they genuinely represent an increased mating effort on the part of the male and are a result of evolutionary pressure on sperm competition.

Unfortunately, as with many forthcoming topics in this book and biology in general, the niceties end here. With sexual conflict rampant in the animal kingdom, it would be naive of biologists to assume that all nuptial gifts were created equal. Indeed, there are cases where gift-giving looks more like *The Grinch Who Stole Christmas* than *Santa Claus is Coming to Town*. Both male and female scorpion flies (family Panorpidae) feed on a rare food source: insect carrion. Males are superior competitors for this hot commodity, and they use it to 'buy' sex from hungry females. To make matters worse, it's common for male flies of many species to try to snatch gifts back from females once sperm transfer has taken place.

Nursery web spider males (*Pisaura mirabilis*) appear to have a few different strategies for successful reproduction. Instead of merely handing over a prey item, males wrap it tightly in silk. It's thought that this extra wrapping increases the handling time of the nuptial gift, hence the length of time a male has to deposit his sperm. You might think this is a legitimate strategy for any male that's providing such a high-quality gift, and you'd be right. It's been shown experimentally that carrying these gifts around while looking for a potential mate has direct (negative) consequences on the speed with which a male can run. What a hero, a knight in shining armour for the ladies of the nursery web spider world! But not all the males are so chivalrous – in approximately 30 per cent of cases, males provide females with dummy gifts: worthless arthropod exoskeletons or plant pieces wrapped up in silk and made to look like the high-quality gifts of their rivals. Since these gifts appear authentic, and come in several layers of wrapping, females must still take the time to unwrap them before discovering their true quality. Although once she's discovered a fake gift a female can abruptly put an end to sperm transfer, there remains a chance (albeit a smaller one) for a male to achieve reproductive success in this way. Since males providing worthless gifts are spending much less time and energy in finding and killing real gifts, they can attempt more copulations. Even the hard-working males that deliver real goods aren't always heroic: male nursery web spiders have been shown to snatch back their gifts once females have terminated sperm transfer.

The lack of chivalry does not end with spiders. Male decorated crickets (*Gryllodes sigillatus*) and katydids (bush crickets, family Tettigoniidae) provide a nuptial gift to females called a spermatophylax. As with many nuptial gifts, it's something for females to eat during the process of sperm transfer; once the female has finished feeding on the spermatophylax, she detaches the sperm-transferring organ and consumes that too – putting an abrupt end to sperm transfer. But unlike the prey-item gifts discussed above, this is a gift that is physiologically synthesized by the male and doesn't seem to benefit the females in any way.

The existence of the spermatophylax is somewhat puzzling to biologists. It's composed largely of water and free amino acids, but studies have shown that females are unlikely to experience any nutritional or physiological rewards for consuming it. Males alone benefit, because they significantly increase their chances of successful fertilization by providing it to females. What's going on here? The amino acid composition of the spermatophylax is a critical factor in determining how much of it will be consumed by the female. In decorated crickets, females discard the spermatophylax and prematurely halt sperm transfer in approximately 25 per cent of cases. When researchers examined the components of both desirable and undesirable gifts, it was found that components that induced a positive 'gustatory' response in females were preferred. In other words, when males made their spermatophylaxes really delicious, females spent more time eating them – regardless of the fact that they were not conferring any nutritional benefit. Affectionately termed the 'candy-maker' hypothesis, this type of behaviour is widespread in all cricket and katydid species that produce spermatophylaxes. It appears the spermatophylax is not so much a gift as a sensory trap designed to maximize sperm transfer.

In some species the ejaculate itself is a form of gift. For reproductive purposes, the only component of the ejaculate that a female requires is the sperm. However, ejaculates are composed of a rich variety of proteins, sugars, water and other substances that females can exploit to their advantage. It's fairly common in many species (including humans) for females to expel ejaculates, and there are many reasons why this may occur. She may be ridding herself of undesirable sperm, she may have been sexually coerced, she may simply not have any further storage capacity, or she may have interest in exploiting it for nutritional content. In several animals from fruit flies to squid, females have been observed to expel ejaculates from their reproductive tracts and consume them. Experiments on beetles have demonstrated that females are more likely to do this when they are deprived of food or water. It's a strategy that makes a lot of biological sense; the female utilizes sex to gain some form of material benefit (aka prostitution in the human world). This demonstrates that both males *and* females have the power to manipulate nuptial gifts to their own advantage.

I'll have what she's having

Remember that teen cult classic movie of the 1980s, *Can't Buy Me Love*, where a very young Patrick Dempsey was turned from zero to hero merely by socializing with the captain of the cheerleading squad? Young Patrick was counting on something that his contemporaries in the film were not: popularity by association.

As social animals, the choices we make about dating/mating partners are open for public observation. A lot of valuable information can be obtained from observing the mating decisions of others – especially those with more mating experience or higher rank. Finding a mate is a process that requires much time and effort, so why not allow someone else to do the legwork for you? For individuals that are young or with little mating experience, copying the choices made by more mature individuals makes a lot of biological sense. Courtship behaviours can be complex and difficult to perform, not to mention difficult to interpret. One cannot necessarily expect to get it right on the first try. However, by carefully examining the strategies of others one can learn to make smart decisions about what makes a good mate. In addition to gaining valuable insight, the copying individuals also save time and energy that can be spent on important tasks such as gathering resources or avoiding predators.

Mate choice copying has been observed across a wide range of ecological strategies and species, from those that exhibit virtually no parental care to those that exhibit a great deal. There does not appear to be one specific set of circumstances under which such a learning strategy could evolve. Indeed, this particular tactic has appeared independently in many animal lineages. From fruit flies to fish, birds, rodents, mammals and primates, the mating choices of one's peers seem to be highly influential.

Humans are not immune to mate choice copying. Both males and females are more inclined to give a high attractiveness rating to members of the opposite sex who are either married or in a long-term partnership. Such ratings are also dependent on the attractiveness of the partners involved. Thus if a man is shown with an unappealing woman, female raters score *his* attractiveness as lower than if he was with a more physically beautiful partner.

Blenny (*Rhabdoblennius nitidus*) and stickleback (*Gasterosteus aculeatus*) (small marine fish species) females take copying to an extreme. They will almost exclusively deposit their eggs into the nest of a male that already has some. For a human female this may seem a ridiculous idea: selecting a male that already has babies by another woman? Far from our world of internal fertilization and massive amounts of female investment, for these fish the maternal workload is small, beginning and ending with egg deposition. So a decision to copy the reproductive choices of other females makes a good deal of biological sense. Males create underwater nests where females deposit their eggs, and they will then guard their brood until they hatch into juveniles. Females can assume that a male chosen by other females is likely to be a suitable sperm donor. Furthermore, a male that has a large number of eggs in his nest is unlikely to desert the brood (improving the likelihood of hatching success and biological fitness for females whose eggs are in his nest) and lastly, males have been observed to more actively court females after they have mated at least once, indicating that they are well aware of their increased reproductive potential once they've got a few eggs in the nest. Extra-'clever' stickleback males have even been observed kidnapping eggs from the nests of other males and tending them in an effort to woo potential females. Others have evolved a specialized structure on their dorsal fin that mimics eggs. Such a structure is intended to trick females into 'believing' that there are eggs in his nest (i.e. that other females have already chosen him).

Female European starlings (birds in the family Sturnidae) provide another fascinating scenario of mate choice copying. Outside the breeding season, females form strong social bonds and roost in close proximity to one another. Such pre-breeding social networks have the power to influence mating decisions among tightly knit female friends. If it proves to be mutually beneficial, females can reduce their own reproductive costs by sharing a mate with a preferred same-sex partner. This is exactly the scenario observed in European starlings (*Sternus vulgaris*). Pre-breeding social partners roost in close proximity to one another and share mates. If additional females are introduced to these nesting sites, the pre-breeding female friends preclude them from mating – like the socially unjust high-school scenario where popular besties share all their resources *and* the captain of the football club.

Mate choice copying in Mexican guppies (*Poecilia mexicana*) is even more bizarre. It calls to mind a timeless and biologically relevant quote from the brilliant Woody Allen: 'Bisexuality immediately doubles your chances for a date on Saturday night.' Small-bodied, cryptically coloured male guppies generally have little success in the mating department. In this species females prefer large-bodied, colourful males, and mate choice copying is prevalent, such that the large colourful suitors often have a harem of females nearby, awaiting their chance at copulation. To gain access to these ladies, subordinate males will initiate a homosexual copulation with the high-ranking male. While this might seem a strange way to solicit heterosexual sex, researchers interested in the ecology of observed homosexual behaviour in male guppies have shown that females mate more frequently with a subordinate male if they had previously observed him initiating a homosexual encounter. It's akin to a human female attending a gay nightclub in search of that sexy heterosexual bachelor she's been looking for all her life. Ridiculous as it may seem for a female to exhibit increased interest in a male once he's been witnessed initiating a homoerotic interlude, the fact that homosexuality is ubiquitous in the animal kingdom charges biologists with the interesting task of discerning its biological relevance.

Is copying behaviour something that is largely confined to the female sex? One might imagine this to be the case, seeing as sexual decisions in females are more energetically important than they are for males, making them choosier. However, we see in many species that mate choice copying is alive and well between males as well. Male guppies have been observed to 'eavesdrop' on the mate choices made by others, mimicking the courtship patterns of conspecific (same species) males even if they are prevented from actually viewing any females. This kind of local enhancement in mating behaviour shows that males will attempt to copulate if another male is doing so – regardless of what the potential target may be.

Male mammals have also been shown to display both eavesdropping behaviour (in the copying individual) and audience effects (in the copied individual). In other words, males alter their behaviour in specific ways depending on whether they are the observing individual, or the individual being observed. In some ways it is a biological conundrum that a male should copy the mating choices of another male where there is internal fertilization and a high degree of parental care. This kind of behaviour increases the risk of sperm competition for both males involved. Why do it? With some animals, such as gazelles, only a small proportion of the female population is sexually receptive at a given time. Female sexual receptivity is not easily detected by males, so for this reason it makes sense for one male to follow the lead of another. There is a huge range in the ability of females to store sperm (from a few hours to several months) and so for some males it could be worth the risk of sperm competition to be able to make a personal deposit.

Once a dominant male has discovered that his mating behaviours are being watched (and potentially copied), he will often alter his actions accordingly. When a male gazelle is alone with two females (one more desirable than the other), he will spend all his time courting the desirable one. However, if he spots an eavesdropping male in the vicinity, he is liable to divide his time equally between the two females – lest he give the eavesdropper some kind of sexual ammunition.

It's not *always* as easy or straightforward as 'dating the popular guy / girl'. Biologists have succeeded in manipulating mating preferences in laboratory settings in order to determine the strength of peer influence when it comes to finding a mate. Individuals that would naturally be in the 'unattractive' or subordinate categories can be made very popular if they are associated with high-ranking partners. Female Caribbean guppies (*Poecilia* species) engage in mate choice copying both in the wild and in the laboratory, showing an overwhelming preference for males associated with other females, regardless of the quality of those males. Sailfin mollie (*Poecilia latipinna*) females take it to another level entirely

– even reversing their own mate choices depending on a male's subsequent interactions with low-quality females. When a highly attractive male is experimentally manipulated into hanging out with unattractive females, observing females will no longer select him as a mate. It's interesting to note how easily females of many species can be 'fooled' into choosing a partner who is clearly of a lower social or biological rank.

Nice guys finish last

When we think about finding a mate, our thoughts often fall to the traditional narrative of boy meets girl, boy wins girl. Many aspects of behaviour in the animal kingdom support this (simplified) view, but there are many times when it couldn't be further from the truth. Generalizations of 'picky females' and 'pushy males' fit in many cases, but the diversity of the natural world means that we can also expect to find plenty of examples that do not.

Not all males are created equal. We human females are well aware this is true of our male counterparts, so it doesn't make sense to assume that adult bird, lizard or mammal males are equal either. Although males often settle their desirability differences through good old honest signalling (by fighting for territories or mates, or by having a superior mating display, song, dance, etc), there are those who resort to alternative tactics that hit below the belt.

A lek mating system is somewhat akin to a bar scene where males come to display their wares to females, and females come to observe said males and to select a mate. Females that attend leks (most research has been done on various bird species) appreciate that they are coming for the sperm deposit, and that's about it. Lekking males rarely contribute anything to the female except their DNA. Something termed 'mating disruption' is a common occurrence in lek breeding birds, and it's exactly what it sounds like: males set out to disrupt the mating attempts of others, competing with each other and in the process potentially increasing their own mating success.

Males of the South American passerine species, the Guinean cock-of-the-rock (*Rupicola rupicola*), engage in two forms of mating disruption. Confrontational disruptions include direct physical contact between two males while one of them is mating with a female. Not only is the copulation interrupted, but the disrupting male may also attempt to mount the female while still in the disrupted male's territory. In non-confrontational disruption, a male will attempt to displace a female from a perch in another male's territory during pre-copulatory courtship rituals of calls and wing-beatings. Disrupted females are likely to engage in a greater number of courtship bouts with more than one male subsequent to being disturbed, perhaps as an assurance of fertilization success after an incomplete mating attempt.

Is there a benefit to this kind of behaviour? It appears there is. Disturbers that focus on specific females during direct confrontations tend to realize the greatest benefits in terms of mating success. In addition to disrupting, they seize the chance to mate with the females that they've disturbed. These females tend to focus their attention away from the male they were with to the male that is currently around (the disrupter), although it's not clear why this should be the case.

Interestingly, the reproductive success of disrupting males is *not* always higher. Having devoted a lot of time and energy to wrecking the romantic efforts of others instead of concentrating on their own, disrupting males can find their reserves depleted when it comes to defending their own territory from jilted would-be suitors or other disturbers, and there may be little time left on the clock for wooing a female the 'old-fashioned way'. Indeed, it is thought that one of the main reasons such mating disruption does not occur in species with short periods of pre-copulatory behaviour is that males who disrupt the mating process for others risk missing out on finding a mate of their own. This is the case for the wood bat, where lekking females visit all males within an area in a limited timeframe. After choosing her favourites, a female will revisit and copulate with them within three to five minutes – which doesn't leave males with enough time to disrupt and court.

Not all males antagonise each other in such a direct way. Male bowerbirds (family Ptilonorhynchidae) engage in a version of mate disruption that's a little sneakier in nature. Bowerbird males are unique in the animal kingdom in that they construct elaborate 'bachelor pads', or bowers, where females come to mate. Aspects of bower construction and decoration are critical to a female's selection of a particular male, and so males put a lot of time and effort into creating a space that will win her affections. In addition to carefully designing and constructing their own bowers, males of all species studied to date have been shown to engage in marauding behaviour, where they attempt to destroy the bowers and steal decorations from other males. Bower destruction by males is a common occurrence; in some species bowers come under attack on a daily basis. When a male MacGregor's bowerbird (*Amblyornis macgregoriae*) is experimentally removed from his bower, it is dismantled and completely destroyed by other males within hours. Those whose bachelor pads have been damaged cannot hope to win the affections of a female until the bower has been repaired, and so the mating success of the destroying male stands to increase during this time. If the destroying male not only wrecks a bower but steals decorations and places them on his own bower, the result is a double loss for the victim, and a double win for the destroyer.

Of course there are trade-offs in bower destruction, because while one is out destroying the bowers of others, one's own bower is left vulnerable. It's been demonstrated that males will form 'stealing' relationships with other males, whereby partners take turns destroying others' love nests. Researchers have determined that optimal marauding strategies vary between bowerbird species according to factors such as bower repair time, male attractiveness and abilities. Despite the risks, males will do what they can to decrease the reproductive success of their rivals.

Fool me once

Deception. When it comes to finding and having sex in the animal kingdom, it's rare when there *isn't* some form of trickery involved. There is great diversity in mating strategies that rely on deceit.

A common form of deception involves females deceiving both their current male partner and other females that are trying to mate with him. In systems where males put a significant effort into parental care, females take measures to ensure that the offspring he's tending to are primarily hers. This isn't an easy task, since in many cases mating is polygamous, with one male mating with several females. For example, in green and black poison frogs (*Dendrobates auratus*), males defend territories and attempt to draw females to mate (and deposit their eggs) within them. Once a substantial courtship ritual (between one male and one female) has taken place, she will oviposit her eggs on his territory. From this point the male will assume complete responsibility for the developing embryos, guarding them and keeping them clean and hydrated. After a few weeks he will transfer the eggs into specifically selected water holes where the developmental transition to tadpoles will occur. In addition to the importance of caring for the developing eggs, selection of an appropriate water hole is critical to survival. Tadpoles of several species are cannibalistic, so a high concentration of individuals within a small area could be extremely bad news for survival of smaller (more recently deposited) eggs.

Thus, in addition to being choosy about with whom she mates, it is in the female's best interest to limit the number of egg clutches from 'other women' so that daddy devotes most of his parental energies to *her* offspring. She can accomplish the latter by being deceptive. Having oviposited in a male's territory, the female will remain nearby, guarding 'her' man; when another receptive female approaches, she will either attack the newcomer or engage the male in a lengthy 'pseudo-courtship' ritual – only to desert him without ovipositing a second time (much to the astonishment and dismay of the male, who would have assumed that more eggs were 'in the bag'). Similar pseudo-courtship behaviour has been observed in a wide variety of amphibian and bird species, and can be applied (albeit loosely) to human behaviour as well.

Females are not the only sex to engage in deception. Males go to great lengths to fool potential partners into the sack. There are many ways in which males attempt to utilize the powers of trickery. A few sections back I talked about the gifts with which males woo females (see 'Trinkets and Tokens'), and how some of these 'gifts' turn out on closer examination to be worthless. In some cases, males engage in even more blatant trickery. For example, there are certain types of gift that females cannot 'cash in' prior to mating. Nutritive products in the seminal fluid can only be enjoyed once the semen has been transferred (i.e. copulation has taken place). Here, it becomes critically important for females to penalize trickster males attempting to give a lower quality gift. In field crickets, where the 'promised gifts' include aspects of the seminal fluid, males can cheat females by providing a lower quality resource. Females retaliate by refusing to re-mate with said males, which lowers (but does not completely diminish) their reproductive success.

Bowerbirds aren't the only species to woo the opposite sex by creating fancy 'bachelor pads'. Bowers are strictly a place for courtship and copulation, *not* for incubation or rearing of young. Consequently, it makes sense for males of many species to create the most 'female-friendly' space that they can, and in most cases they do just that. To overcome the fact that more successful males build better nests, inferior males resort to deception in order to make their second-rate offerings more alluring. For example, among cichlid fish (family Cichlidae) in Lake Malawi bower height is an important factor, signalling a male's superiority to other males and hence making him more desirable to females. Some inferior males give themselves a head start in creating a taller tower by constructing their bowers on rock-platforms instead of the sandy lakebed. This not only increases the bower's height but also reduces costs in terms of construction, maintenance and competition. Although in the human world we might see this as an act of ingenuity, in the animal world it serves as an unreliable sign of male investment (i.e. a deceptive trap). Interestingly, though females are more likely to visit the tall, deceptive bowers, the majority looks them over and moves on. Since their lazy efforts cannot withstand close scrutiny, deceptive males are not winning the battle.

In addition to creating elaborate bowers, male great bowerbirds create optical illusions to make their signals even more enticing to potential female partners. 'Forced perspective' is an illusion created by arranging objects in such a way that they increase in size as distance from the bower entrance increases. The quality of the illusion is highly correlated with reproductive success, indicating that females choose the males who fool them the most. In fairness to great bower females, only the most intelligent males can successfully create such an illusion, so in this case the foolery is actually representative of an honest biology. Are we all thoroughly confused now?

There are other ways that males can deceive females into copulating with them. Male topi antelopes (*Damaliscus korrigum*) manipulate females into staying on their territories by tapping into the power of fear. When females attempt to exit the territory of a certain male, he will make fake alarm-call vocalizations, signalling to the female that she's headed directly into the clutches of a predator. This strategy is quite effective in keeping females around for further mating opportunities – although one does wonder how many times a specific female will fall for the ruse.

Peacocks (*Pavo cristatus*) also utilize the deceptive power of vocalizations, although in this case the male puts on a fake sex show to lure unsuspecting females. Peahens will often select a male based on the choices of others (see 'I'll have what she's having'). Therefore, if a male can deceive a female into believing that he's been chosen (in this case by making vocalizations to give the impression he's copulating in the bushes), it can have the effect of drawing more females in for a visit. Manipulating vocalizations to fool a potential mate into copulation is a fairly basic strategy that can work – provided it's not used too often. That's the key to any form of deceptive action: use it too often and it becomes expected – and ineffective.

I've touched on females deceiving males, and males deceiving females . . . how about males and females working together to deceive? Gelada baboons (*Theropithecus gelada*) live in large social groups where there is typically one dominant (alpha) male. The dominant male gets most of the sexual action, much to the chagrin of other sexually mature males in the group. However, subdominant males can still partake in some sexual activity, provided they can successfully deceive the alpha male. Individual males and females that engage in 'forbidden' copulations do so in sneaky fashion, partaking only when the alpha male is some distance away. In addition, both male and female cheaters suppress vocalizations associated with sexual intercourse, sharing their forbidden passion in a blanket of silence. It is important for the cheaters to be careful – if the cuckolded (alpha) male were to become aware of their affair there's a risk he would use his strength to punish the guilty parties.

In all the examples I've discussed so far, regardless of who is being deceitful, we can consider one party to be the winner and the other to be the loser. At least one individual benefits through the deceit by increasing their individual biological fitness. That's not always the case though. Sometimes, males and females lose out simultaneously. Such is the case when sexual deceit comes from an entirely different species.

A sexually parasitic strategy is employed by blister beetle (*Meloe franciscanus*) larvae, which deceive both sexes of a species of solitary bee (*Habropoda pallida*). In this case, aggregations of larvae mimic the chemical and visual signals of female bees. Males are lured in by these signals, and upon copulation the parasitic larvae are transferred to his body. When a male does find a sexually receptive female of his own species, the parasites are transferred to her body during copulation. They will then be transferred to her nest as she oviposits her newly fertilized embryos. The blister beetle larvae feed on eggs and pollen of their female bee hosts, resulting in a 0 per cent chance of bee survival for any infected nest, and a diminished biological fitness for both male and female bees.

By spending a substantial amount of energy, time and sperm on someone else's reproductive success, individuals involved in this kind of deceptive courtship and copulation are indirectly decreasing their own. Other species do not stop at parasitism, but use sex as a murder weapon. Many predators utilize the power of sexual deception to lure unsuspecting bachelors to their death. Bolas spiders (*Mastophora* species) have evolved the ability to mimic the sex pheromones of female moths such that males seeking out a receptive female find themselves caught in a sticky trap of spider silk. So sophisticated is the mimicry of the Bolas spider that they will go 'fishing' for different moth species at different times of night, depending on when each moth is most active.

Predatory katydids (*Chlorobalius leucovirdis*) mimic the acoustic and visual sexual signals of female cicadas by imitating species-specific wing-flick songs (characteristic sound patterns created by flipping the wings in a particular way), and by making synchronized body movements. Female *Photuris* fireflies hunt by mimicking the specific light-flash patterns of their prey species (also fireflies). Interestingly, male photurids will mimic the light-flick patterns of the prey species, so that they increase their chances of coming into contact with a sexually receptive female! Yes, you read that right. She's hunting, he's posing as the prey.

Secrets and lies

As we have seen, sexual deception is a biological bummer, at least for the deceived party. However, we have yet to meet some of the most blatantly sexually deceitful beings on the planet. They engage in the ultimate in exploitation. I'm talking about sexually deceptive plants, and all the poor arthropod males who are wasting their reproductive efforts on entities that aren't even animal. The majority of examples of sexual deception come from members of the orchid family (Orchidaceae), although examples from an iris and a daisy have recently been discovered, indicating that sexual deception is probably widespread over the entire planet. It has evolved independently on at least four continents, and its existence continues to perplex evolutionary biologists and botanists alike. An initial question springs to mind: why would a plant want an animal to have sex with it? Plants require animals (or other interventions) to cross-pollinate, and sexual deception is thought to have evolved from plants that offer food rewards to pollinating insects. Except, in this instance, instead of receiving a food gift, the duped male suitor receives absolutely nothing.

Most systems of sexual deceit are species-specific: a single insect species, for example, will pollinate a single orchid species.

This is in direct contrast to the systems of food-gifting pollinators, which may have a wide variety of pollinating insects helping to spread their seeds. It would appear that having a single pollinator is more effective, since less pollen is lost or deposited on the wrong species.

So how do these plants so effectively entice the attention of would-be male suitors? The most definitive answer is in the chemistry. Over millions of years orchids have evolved countless chemicals that are exact replicas of reproductively active female arthropods (most generally members of the Hymenoptera, one of the largest insect orders containing bees and wasps). The accuracy of this chemical mimicry is astounding, and this is undoubtedly what draws the male's attention at the outset, but an effective female disguise must continue the foolery even when the male is close enough for his vision to come into play.

We see remarkable structures and colours on many orchid species, which run the gamut from simple yellow pigmentation (a wavelength that is most discernible within the visual spectrum of the hymenoptera) of the pollen, to bright yellow sepals and petals. The shape of the flower, as well as its UV reflectance and size are important factors for males of different hymenoptera. In one orchid species, the UV reflectance of bumps on its labellum (or inner structure) is a perfect match to the UV reflectance found on the wings of sexually receptive females. In extreme cases the labellum may have evolved to appear as an exact replica of the female in question, and in some species it has evolved to be a slightly larger version of a sexually receptive female in order to exploit the male's natural preference for females with a larger body size.

As if the deck wasn't stacked enough against male hymenopterans, there are also several factors that render specific hymenopteran species particularly susceptible to deception.

Solitary species (as opposed to colonial species) are liable to be sexually deceived because males have evolved mechanisms to be attracted to airborne sex chemicals from potential female partners. They engage in intense mate searching and tend to show fairly robust mating behaviours once females are located. While males are polygynous (meaning they will mate with lots of females), the females are obligately monandrous, i.e. the first sperm a female receives will sire the majority (or all) of her offspring. These ecological characteristics provide the perfect scenario for deceptive orchids: males who are easily drawn by chemicals, and try to fertilize as many virgin females as they possibly can. In addition, the majority of species subject to sexual deception are haplodiploid – which means that fertilized eggs produce female offspring, and unfertilized eggs produce male offspring. It's a reproductive strategy that allows females to produce offspring (males) in the absence of a sperm donor, and it's an important aspect of the solitary nature of these species.

Once a male pollinator has located his orchid lover, he will engage in a variety of behaviours, again dependent on the species. These behaviours range from briefly touching down to full-out sexual action and ejaculation. The delightful term pseudocopulation encompasses the acts of intense sexual behaviour that many males employ on their 'females'. Males will erect their genital capsules and make convulsive movements, rubbing them against various parts of the flower. In some cases, the labellum (or inner part of the orchid flower) acts as a spring trap, responding to male stimulation by closing up and temporarily trapping the male inside. In an extreme case involving members of the orchid genus *Pterostylis*, this touch-sensitive trap snaps shut in response to the probing sexual behaviour of a male gnat. The orchid doesn't immediately open, forcing the male to exit through a small tunnel that brings him into direct contact with the plant's reproductive structures.

This seems an energetically costly process for the orchid (not to mention for the gnat, who must manoeuvre through its genitals to find the exit), but it recovers after three hours, and has the ability to be 'sprung' three times during its reproductive life.

So it seems that, in this case and several others, there is the possibility for significant costs to the male. There are examples where orchid deception is so accurate that males prefer to copulate with a flower rather than a true female. In other cases a male may prematurely end copulation with a true female in order to visit an orchid. Such 'wrong decisions' prove costly to a male in the context of biological fitness, as do cases of mistaken identity when the density of mimicking orchids is high in a particular area. Once again, all males are not created equal, nor are all orchids. Evolution favours orchids with the highest level of sexual deception, and favours male Hymenopterans with the most acute ability to discern them from females. Although biologists have long assumed that the fitness consequences of sexual deception on male hymenoptera are negligible due to the low cost of sperm, it has recently been argued that the loss of time and energy incurred by fooled males could be substantial. Male pollinating bees that are the targets of the sexually deceptive African daisy (*Osteospermum* species) exhibit a distinct learning behaviour over the short term; if they have visited a daisy (as opposed to a true female), they are less likely to visit another one immediately afterwards. This avoidance behaviour is critical because it reflects the fact that there must be some kind of cost to the male for engaging with the daisy in the first place. It implies a form of antagonistic coevolution between sexual deceiver and sexual fool. Unfortunately, male Hymenopterans seem to have short memories; after twenty-four hours they cease to 'remember' that orchids are not sexual partners. This may be critical to the maintenance of sexual deception overall.

Another example that demonstrates a significant male cost is that of Australian tongue orchids (*Cryptostylis* species) and their fooled pollinators, the orchid dupe wasps (*Lissopimpla excels*). Here, male wasps gain such a high level of sexual arousal from the orchids that it ultimately results in ejaculation. Copious amounts of sperm are wasted this way, which clearly represents a cost to the spent males, who can experience refractory periods prior to being able to ejaculate again. Here's where the story takes an interesting twist: we would expect, based on the high cost incurred by the males of ejaculating in plants, that this strategy would result in a lower pollination success for the orchid. In other words, since it's negative for the male pollinator, we'd expect him to evolve mechanisms to avoid having sex with the orchids, thus leading to a lower rate of overall pollination. In fact, the opposite is true. Pollination success of the tongue orchid is among the highest observed for any sexually deceptive species. What is going on here? It may have a lot to do with the haplodiploid mating system employed by the Hymenopterans. Even if males end up 'spent' after mating with orchids, unfertilized females can still produce more males . . . who will also be naive with respect to such sordid sex. In addition, evolutionary selection on males is tricky for haplodiploids because it can only happen indirectly through daughters, then through their sons. It's possible that the lack of sexual reproduction in the dupe wasps, leading to the production of greater numbers of naive males who will continue the cycle, is beneficial to the orchid. As long as a few males manage to fertilize females to maintain genetic diversity (and females) in the population, there's nothing to hinder the evolution of a system in which a good amount of reproductive effort is wasted on orchids.

You may be wondering, does sexual deception by plants always manipulate the male of a species? In the vast majority of cases, the answer is yes; but there is a subset of sexual deception called 'brood-site deception' that acts on females. This form of deception involves the female's part of the sex act: oviposition. It's arguable whether this can truly be classified as sexual deception since it doesn't occur until successful copulation between male and female arthropods has taken place, but it still interferes in the reproductive process so I'm going to classify it as such. Brood-site deceivers (orchids that are generally pollinated by flies or beetles) manipulate their chemistry and morphology to mimic places where females would normally deposit their eggs. Female flies and beetles are not known for ovipositing in the nicest spots; common sites include carrion, dung, fermenting fruit, yeast or fungi. Several species across many flower families have evolved to emit sulphur-based chemicals that mimic animal dung or carrion. The orchid *Dracula chesteronii* has evolved to morphologically resemble the gill cap of a mushroom, and the orchid *Paphiopedilum dianthum* has evolved tiny, contrasting spots that mimic the aphid prey of its pollinators. The helmet orchid (*Pterostylis* species) resembles the wild mushroom oviposition site of the fungus gnat (*Mycetophilidae*) in both colour and UV reflectance. Other species that mimic the dull colour of dung or the bright red of carrion show that the evolutionary process can work in some remarkably disgusting ways.

Sex or sick?

When it comes to finding a mate, humans focus on the positive and aim to put our best foot forward. We embellish our online dating profiles with flattering selfies, we preen ourselves in readiness for dates, and we try to engage in pleasant and witty conversation. There will be periods when circumstances are against us – if we have flu, for example, or we've broken a bone – but since we aren't seasonal breeders, we don't need to worry about missing out on a year's worth of reproductive opportunity because of a poorly timed bout of illness, and most of us have access to medical supplies and a bed to which we can retreat. This is not the case for species who have to make time-critical decisions about finding a partner and investing in reproduction with them. In the animal kingdom, mate choice is dependent on condition, so wounded or unhealthy individuals will find themselves at a major disadvantage in the mating game. This isn't just because they are less attractive to the opposite sex (although that's certainly part of it). If they are sick they won't have the stamina for a prolonged search for a mate, the odds will be stacked against them in intrasexual competition for sexual partners, and even if they do manage to win the affections of an unsuspecting member of the opposite sex, chances are that the partner is not going to stick around once their degenerate condition becomes apparent.

Distinctive 'condition categories' between individuals aren't always a matter of sickness or injury. The difference in quality may be a simple function of genetics: low-quality parents produce low-quality offspring, who go on to produce more low-quality offspring with other low-quality partners. But it is possible to alter an individual's quality by manipulating the conditions in which they are reared. Zebra finch (*Taeniopygia guttata*) offspring born into small clutches get more parental care, food and nurturing, and can be reliably assumed to be higher quality. However, by increasing the size of a brood, biologists found they could produce low-quality offspring.

Interestingly, female zebra finch offspring of both high and low quality exhibit 'condition dependent' mating choice, i.e. they select partners from within their own quality category. Now, why wouldn't any female 'aim for the stars' when it comes to mate selection? After all, if there's one thing that biologists harp on about, it's that eggs are expensive and so a female should be choosy about the males she shares them with. The answer lies in the fact that zebra finches (like many birds) exhibit bi-parental care, which means the male investment is substantial, and so there is mutual mate choice. So in this instance it isn't simply a matter of the choosy female and the desperate, sperm-flinging male. Thanks to male choosiness and competition from higher-quality females, a low-quality zebra finch female probably wouldn't stand a chance of mating with a high-quality male. A similar level of female choosiness (or lack thereof) is observed when female quality is altered later in life. When researchers clipped the flight feathers of female canaries (thereby decreasing their flight ability and attractiveness) it was found to decrease their level of choosiness when selecting a mating partner.

What about when a perfectly healthy (high-quality) individual gets hurt? After all, there are many animal species that are aggressive towards each other, or to other species nearby, or who live in areas with treacherous terrain that can lead to injuries or wounds. Depending on several aspects of the ecology of the species in question, it seems injuries can have a drastic impact on an individual's chances of reproductive success. Tree lizards (*Urosaurus ornatus*) are an aggressive species that frequently engage in brawls with each other. Indeed, individuals in the wild are often found with wounds on their external surfaces. When subjected to experimental wounds ('subcutaneous' biopsies), researchers found that female tree lizards invest more in immune response than in reproduction. This was measured by counting the number and size of egg follicles within their bodies, and the level of reproductive hormones in the blood. When females were provided with unlimited food, they were able to allocate energy to both healing and reproduction; however, in the absence of food, females did not allocate energy to reproduction. That is hardly surprising: when one has a wound to heal or a broken bone to repair, energy should be allocated to getting better rather than getting lucky. But what happens when injuries are more severe, or one has been infected by a parasite or virus that's only going to get worse?

If individuals are young and otherwise healthy, scientific studies suggest it's more likely that a slow-down of reproductive effort will ensue. With every reason to suppose that future reproductive opportunities will arise, it makes sense for young (otherwise healthy) individuals to get better before getting busy. However, when an illness is critical, or when infected individuals are older, it may make more sense to increase reproductive effort – sometimes to irreversible health detriment or even death – in a last-ditch attempt at passing on genetic blueprints to future generations. This phenomenon is called 'terminal investment' and it illustrates the concept of biological fitness in the face of illness. Alterations in reproductive ecology due to pathogenic infection have been observed in many organisms, with changes to rates of oviposition, earlier maturation (but at a smaller size), changes to the size or number of eggs produced, rate of pheromone production, quality of secondary sexual characteristics, and quality of ejaculate. Terminal investment presents a fascinating case study in the evolution of pathogenic infection because terminally investing individuals can give out dishonest reproductive signals.

Male crickets (*Allonemobius socius*) afflicted with a bacterial infection redouble their reproductive efforts by making their songs more energetic (and therefore more attractive to females). Having used up their reserves of energy on the song, such males cannot proffer a high-quality nuptial gift. Their energetic song is therefore a dishonest reproductive signal that presents a risk to swooning females, who are deprived of precious parental resources by his low-quality gift and run the risk of contracting the infection. Females that mate with such infected males are liable to experience a decrease in fitness, which means that this kind of dishonest signalling is unlikely to become prevalent in a population (though it can evolve to occur in low numbers).

Sperm quality is another honest signal that is affected by immune challenges. The body often identifies spermatozoa as 'foreign' objects, but they are protected from immunological attack through immunosuppression. However, when the body is under siege from a bacterial infection additional immune responses are mobilized which could result in an attack on a male's own sperm, reducing ejaculate quality. This is precisely what happens to male crickets and other arthropods when experimentally afflicted with bacterial infections.

Male leopard frogs (*Rana* species) infected with amphibian Chytrid fungus (*Batrachochytrium dendrobatidis*), a disease that has been linked to amphibian declines and extinctions worldwide, experience an increase in the width and volume of their testes, along with an increase in sperm production. This could be interpreted as an honest signal of terminal sexual investment, but biologists have yet to determine the quality of the sperm. It cannot be classified as an honest terminal investment if the infection has resulted in a decrease in sperm quality.

These examples make me truly thankful to be human, for we are able to cure the vast majority of our maladies prior to getting busy in the bedroom. Although we are not without our own species-specific versions of terminal investment (e.g. harvesting eggs or sperm prior to a patient undergoing chemotherapy or radiation), these are nowhere near as prevalent as in the natural world.

If we've learned anything from the chapters in the preceding section, it's that the process of sex begins long before gametes are exchanged. The ability to recognize and successfully court an appropriate partner is paramount to the ultimate goal of copulation, and it's rarely a straightforward operation. Social, biological and environmental factors can all affect the outcome of mate searching and courtship, and at times it may seem like a near impossibility that any animal manages to find an appropriate partner at an appropriate time. Despite the risks of disease, deception, and downright dirty tricks, the bottom line is that for most members of any given species the obstacles can generally be overcome for at least one bout of prospective partnering. Now that said, partners are in place, it's time for the next critical phase: the actual sex.

SECTION TWO

THE SEX

It's so easy for humans to forget that our own attempts at finding a mate are blissfully free from the many distractions faced by our animal cousins. As I mention above, finding a mate is not without its difficulties. However, once the difficulties have been overcome and partners have been located and appropriately courted, the next step is to engage in actual sexual deeds. For members of our species, this is where the fun begins. For everyone else in the animal kingdom: not so much. Despite the fact that copulation is imminent, there remain a multitude of factors that can alter the success of a particular coital event.

Ejaculate!

When we think about sex between males and females, we generally view the male's role as fairly straightforward: get sperm to egg. Full stop. Although males are often only responsible for the transfer of one molecule, it is not necessarily a straightforward process even once sex has taken place. It is truly misleading to think of copulation as mere sperm delivery, because ejaculates are actually very complex. Hundreds of active components are present in the seminal fluid, including proteins, fats, salts and sugars. There is a myriad of circumstances that influence whether the seed of a particular male will be the successful fertilizer, and a number of ways that a male can increase his success. Specifically, he can create a delectable concoction of proteins and other macromolecules that serves various purposes to meet his needs. This rich selection is mainly produced by accessory sex glands in both invertebrates and vertebrates. One can imagine that such an array of goodies has a variety of functions – and indeed it does.

Protein factors have been identified from ejaculates across a wide variety of animals that play roles in activating sperm once it has reached the female reproductive tract and that function to increase its motility. The ability to swim more quickly is particularly important for organisms in which a female mates with several males in rapid succession or where the female reproductive tract is especially hostile. Vaginas tend to be acidic, and low pH has a significantly negative effect on the ability of sperm to swim. In addition to compounds that increase swimming speed, many ejaculates contain bicarbonate to keep the sperm in a more alkaline environment as it makes its way to the eggs. To defend against the female's immune system mounting an attack on spermatozoa – which it identifies as 'foreign bodies' (as indeed they are) – ejaculates usually contain components to impede any such assault. Substances are produced that not only protect sperms by forming a defensive barrier around them, but also act to 'corral' them into areas of the female's reproductive tract (e.g. her spermatheca) where they will not be attacked. Sperm may also be subject to oxidative damage from free radicals in the female's reproductive tract, so ejaculates often contain 'superoxide anion scavengers' like vitamins C and E.

These various compounds serve to increase the success of a male's spawn once he's managed to plant it in the right spot; but there are other compounds that serve to make the sperm of *other* males less desirable. Allospermicidal components of ejaculates (those that have negative effects on the sperm of others) can act to remove sperm that has been previously deposited in a female's reproductive tract. Still other proteins (such as those secreted from the male accessory gland in fruit flies) have evolved to defend one's sperm against such nastiness.

The complex array of manipulative substances does not end with effects on the ejaculates themselves (although the positive effects for males are certainly a large part of it). There are also substances that effect changes in the females who receive them, all in the name of increased reproductive success. Several compounds have been identified that increase a female's short-term investment in reproduction, including those that function to increase the number of eggs produced by a female, the size of those eggs, and the act of oviposition. In the Mediterranean flour moth (*Ephestia kuehniella*), compounds within the seminal fluid signal females to allocate more resources to eggs, while the sperm themselves signal females to lay them. While this seems like good news both for males (who increase fitness by successfully fertilizing females) and for females (who increase fitness by laying more eggs), such compounds have overall negative effects on females' longevity. In many species where males stimulate a short-term increase in female fecundity, the ultimate result for her is a shorter lifespan.

Other compounds within the seminal fluid have different effects on females. For example, instead of ramping up egg production and laying, they may be induced into a refractory (sexual disinterest) phase. Male bumblebee (*Bombus* species) ejaculates contain cycloprolylproline, a substance that decreases the frequency of female re-mating, and those of male *Drosophila* (fruit flies) contain peptides that serve to decrease female receptivity to male sexual advances. Male corn-earworm moths (*Helicoverpa zea*) have ejaculates that contain a pheromonostatic peptide, which has the short-term effect of depleting the pheromones of females (hence making them unattractive to potential male partners).

In addition to containing compounds that affect a female's behaviour, ejaculates can also contain substances with coagulative properties that 'seal up' a female's vaginal opening once a male has made his sexual deposit in order to decrease the ease with which she can re-mate. Fatty acids (linoleic, palmitic, oleic and stearic) contribute to the formation of mating plugs in several invertebrate species, while a vast array of proteins and peptides contribute to seminal coagulation in vertebrates. For primates, the level of seminal coagulation is most pronounced in species where female mates multiply. This is but one example of how reproductive ecology can shape the evolution of ejaculate composition.

While it may seem horribly unfair for a female to be subjected to a suite of compounds designed to manipulate her immune system, her reproductive system, her post-coital behaviour and the status of her vaginal opening, there are some components of a male's ejaculate that provide distinct advantages to females. For example, there is a good deal of nutritional value in a typical ejaculate. Sugars, salts and fluids can be utilized by females in various ways. In the beetle *Bruchidius dorsalis*, males transfer huge ejaculates to females: approximately 7 per cent of their body weight. This substantial size can have direct (i.e. nutritional) and indirect (i.e. increased fecundity due to higher nutritional status) benefits to females. Often the current nutritional or hydration state of a female will dictate whether she seeks additional copulations. For example, when female seed beetles (*Callosobruchus maculatus*) are experimentally deprived of food and water, they will seek extra copulations with males in order to obtain said resources. If they are well fed, such 'superfluous' sex does not take place.

Females can also benefit from immunostimulatory properties of the seminal fluid. Ejaculates often contain compounds that have antibacterial, antifungal or antiviral properties designed to minimize microbial damage to sperm. Once inside the female reproductive tract they can provide the same services for the female, essentially providing her with extra defences to protect against foreign invaders. In some cases males also contribute probiotic bacteria such as *Lactobacillus* that can benefit females by out-competing more pathogenic bacteria.

Far from simply being a way for sperm to meet egg (although that's the ultimate goal), the various components of seminal fluid are an ecosystem of their own, which has to respond to the various demands of a sexual system. To this end, one might imagine that evolution would select for plasticity in ejaculate content, since a male will not always find himself in identical circumstances when it comes to his sexual exploits. Indeed, ecological context is an extremely important aspect of ejaculate composition. Males appear to vary their concoctions according to the risk of sperm competition. For example, if a male is mating with a virgin female (i.e. no sperm competition) it makes more sense for him to allocate more resources to components of the semen that will induce egg production and laying in his female partner rather than providing her with large amounts of sperm. There is evidence to support this in some invertebrate species. For example, male fruit flies (*Drosophila melanogaster*) deliver peptides in their ejaculates that stimulate egg production following mating and regulate the maturation of eggs. Males appear to be sensitive to the mating history of female partners in that the levels of these components varies according to whether or not they are mating with virgins. When females have already mated (and hence have received a dose of ovulin – the seminal protein that stimulates fecundity), males are less likely to include high concentrations of it in their ejaculates. Why waste precious energy providing her with a compound that's not going to have any measurable effect?

Mating status isn't the only factor that a male may consider as he's concocting his ejaculatory potion. Nutritional status, rank, age and condition are all aspects of mate selection that could influence the types of compounds that will prove advantageous. After all, females will vary in their responses to fecundity stimulation or receptivity inhibition depending on their current status. In addition, if a female is of low quality a male may be less inclined to invest in an expensive ejaculate; however, if she is of high quality a male will be required to invest to a greater degree because he's liable to face a higher level of sperm competition in mating with her. Accordingly, male domestic fowl (*Gallus gallus*) transfer larger ejaculates with high levels of compounds to increase sperm velocity to particularly attractive females.

Of course it's not just females that can vary in quality. Males of various species exhibit a wide range of variation in their rank, health and nutritional status. A male who is in poor physiological condition will not have the ability to provide 'expensive' components in his ejaculate that could increase his reproductive success (refractory components, for example, are especially costly). Instead, low-quality males often produce ejaculates that are high in water content, as they quickly run out of protein or lipid-based compounds. In addition, low-quality ejaculates stand to be out-competed by ejaculates from high-quality males, and low-quality males take longer to replenish seminal fluids and sperm subsequent to copulation. In this way, ejaculate composition is a true reflection of a male's biological fitness. The story isn't as simple as that though. High-quality males (in theory) have the ability to produce top-quality ejaculates. They are also in demand among females, meaning that a five-star male could get tapped out if he's too generous with his donations. Indeed, high-quality male tree crickets (*Oecanthus nigricornis*) lower the value of their ejaculates so as to be able to 'serve' a larger number of ladies. It may seem a bit of a dirty trick – manipulating ejaculate quality according to the number of willing female participants – but it's not the only dirty trick playing out when it comes to components of the seminal fluid.

Something called 'ejaculate exploitation' is fairly common in polyandrous mating systems (where a female mates with multiple males). Males can take advantage of the ejaculates that have come before theirs in order to maximize their reproductive success at a minimal cost. In other words, if multiple matings take place in a short amount of time, chances are that some of the seminal fluids from the previous partner will remain in a female's reproductive tract. Male fruit flies take advantage of the fact that previous ejaculates contain compounds that protect sperm within the female's reproductive tract, producing seminal fluid with a lower concentration of those protective compounds. It makes the entire process more efficient for the later-mating males, who can enjoy the benefits of sperm protection that's already present. Why add more of the same compounds if they are in situ and doing their intended job?

Another example of ejaculate exploitation comes from the externally fertilizing grass goby fish (*Zosterisessor ophiocephalus*), where there are two distinct male strategies: territorial and sneaker. The territorial male guards an area where a female will lay her eggs, and he fertilizes them once they have been deposited. The sneaker male tends to be much smaller and more cryptic (camouflaged) than his territorial counterparts, and he will achieve a good deal of reproductive success by hiding nearby and darting out to fertilize some of the eggs at the exact moment the territorial male fertilizes. Ejaculates of territorial versus sneaker males vary greatly in their composition. Seminal fluid of territorial males is viscous and abundant, which helps to protect the sperm in the seawater during the eight-hour period of spawning. For this reason, their ejaculates are high in seminal fluid and low (relatively speaking) in sperm content compared to those of sneakers. They also contain substances that enhance the sperm's mobility, which makes sense considering they must travel through a fairly inhospitable environment prior to fertilization. Sneaker males take advantage of the ejaculate properties of their territorial rivals, specifically by making use of the speed-enhancing compounds. In this way the sneakers exploit the benefits of high-quality seminal fluids without having to spend their own energy in producing them. In fact, when the sperm of territorial and sneaker males is combined, that of the sneakers actually travels faster. The seminal fluid of sneakers appears to have negative effects on the sperm of the territorials, although the exact mechanism for this remains unknown.

May the best sperm win

As we learned in the previous section, not all ejaculates are created equal. Well, not all sperms are created equal either – there is a great deal of variation in many aspects of sperm quality including morphology, mobility and viability. As with all other aspects of biodiversity, there is tremendous selective pressure at the cellular level, especially with respect to the gametes. Males produce hundreds of millions of sperms, and so there is competition both within and between individuals as to which one will ultimately be the successful seed. There are a few basic generalizations that can be made about the characteristics that categorize a successful sperm, although as usual there are exceptions. First, sperm that are under intense selection pressure tend to swim faster than ones that are not. In other words, the sperm of organisms with highly promiscuous mating systems is generally faster than that of organisms that are monogamous. Similarly, the number of sperms per ejaculate tends to be higher in promiscuous organisms as opposed to monogamous ones. These generalizations make a lot of sense in terms of the various battles that an individual sperm may face; however, there are many aspects of sperm competition that don't immediately come to mind. Different mating strategies require different adaptations of sperm.

Consider organisms that are external versus internal fertilizers. As humans, we're prone to forget that there are many organisms that spew their gametes into the outside world rather than into the discrete genital openings of their partners. Many aquatic organisms such as fish, invertebrates and amphibians adopt this method, utilizing their watery environment to buffer the effects of the elements on their delicate gametes. The sperm of fish species often follows the general pattern of being faster in those with promiscuous mating systems. Among the closely related cichlid species (family Cichlidae) inhabiting Lake Tanganyika, those with promiscuous mating systems not only have faster-swimming sperm, but higher numbers per ejaculate, larger size, and greater longevity. Sperm swimming speed is a critical factor for external fertilizers, so we naturally expect this characteristic to be under a lot of selective pressure. Interestingly, the situation with externally fertilizing frog species is quite different – it's the slower-swimming sperm that are usually more successful. While this initially seems counter-intuitive, in the context of frog ecology it makes perfect sense.

Frog sperm have a longer, more difficult journey to their targets than those of most other aquatic organisms. Eggs are encased in several thick layers of viscous jelly, and the majority of sperm fail to reach them after becoming trapped within the gelatinous matrix. Fast-swimming sperm are less successful in this scenario because they run out of energy prior to reaching the eggs. Slow and steady sperm gain the greatest level of success in navigating the course. Reaching the eggs is the *most* critical thing for externally fertilizing fish and frogs; however, the sperm of organisms with internal fertilization have other factors to consider. Many females can store sperm for long periods in structures called spermathecas, which sets the scene for a good deal of sperm competition to take place within these storage organs.

How should a male alter his sperm characteristics in order to be a successful competitor? When females have the ability to preferentially select certain spermatophores (or sperm packages), the game changes remarkably. In some cases, less is more. For example, male springtails (*Collembola* species) decrease the number of sperm-packages that they create when there is an increased incidence of competition from other males. In these organisms, sperm transfer is dissociated, meaning that males simply drop off their spermatophores in the environment, and at a later time females will come and pick them up. It's like leaving a letter in a mailbox that a female may or may not want to read. In this way, females have the power to choose spermatophores outside of direct pressure from males. When male:male competition is scarce, it makes sense for males to create as many (low-quality) spermatophores as possible in order to reach a maximum number of females. Conversely, when males have to compete for a female's attention they invest a greater amount of energy into creating high-quality sperm deposits. Indeed, females exhibit a preference for spermatophores from males that have been exposed to a competitor, indicating that a little competition causes males to produce sperm of higher quality.

Females with sperm-storage organs throw an additional complicating factor into the mix because their ability to store sperm allows them to collect donations from several suitors and let them compete for a while within the storage organs prior to fertilization. These delays ensure that only the fittest sperm will thrive. It's a strategy that can be observed in organisms from diverse phyla. Consider the ecology of honeybees (*Apis* species), where virgin queens take their nuptial flight and receive sperm donations from between ten and twenty drones. This will be her one and only chance to collect sperm, and she will then spend the rest of her life fertilizing her eggs from her home in the hive. In that same vein, male drones exist for the sole purpose of mating with the queen on her nuptial flight – after which they die. You can well imagine that the selection pressure on sperm quality is extremely high in these and other eusocial insects (highly cooperative and organized societies like ants and bees). Queens will discard more than 90 per cent of the ejaculate that they receive; only around 2.5 per cent of it will ever migrate to long-term storage. This means that a drone's chances of reproduction are devastatingly low. The only way to increase them is to produce sperm of high quality, with viability and stamina.

The sperm of male desert ants (*Cataglyphis savignyi*), another eusocial species where selection pressure on sperm quality is also significantly high, has evolved a group-swimming strategy in order to attain a high level of reproductive success. Sperm from a given male will aggregate into bundles of 50–100 spermatozoa called spermatodesmata. These bundles have a common head-structure that keeps the sperm altogether, and they receive a respectable increase in swimming speed. When compared to solitary sperm cells, sperm in spermatodesmata swim 51 per cent faster, and clearly have an advantage when it comes to reaching a valuable space inside of a queen's spermatheca. Charge!

There are a few other scenarios that we should consider where sperm storage is concerned. First, many organisms exhibit a 'last male' advantage, meaning that the sperm of the most recent suitor fertilizes the largest proportion of a female's eggs. In this kind of scenario the *timing* of copulation is as critical as the quality of sperm being delivered. On the other hand, there are organisms that follow a 'fair raffle' strategy when it comes to sperm storage. All sperm are mixed together within the spermatheca regardless of mating order.

Prolonged sperm storage is especially advantageous for organisms that have low encounter rates with the opposite sex, as is the case in tortoises where females can store sperm for up to four years. This is an extremely long period of storage, made possible by the evolution of specific sperm storage tubules within the female's reproductive tract. In fact, many bird and amphibian species have evolved such tubules and utilize a fair raffle strategy. What can a male do to maximize his sperm's chance of being competitively successful in long-term storage? Make lots of them! Indeed, when competition is high, males stand to benefit more if they can pump out enormous numbers of sperm to eke out a corner in the storage space. Although rare in mammals, many bat species employ sperm-storage as a reproductive strategy. Where this is the case, males have significantly larger testes than in bat species where sperm storage doesn't occur. When your sperm is going to be subjected to long-term competition it's best to ensure substantial quantities.

Aside from the overall generalizations of sperm characteristics that I've been discussing so far, there are also scenarios where males vary their sperm characteristics in accordance with the breeding season or other ecological factors. After all, maintenance of high-quality sperm is metabolically costly. If it's not going to be put to immediate use, it makes much more sense for a male to spend his valuable energy elsewhere. Red-winged blackbirds (*Agelaius phoeniceus*) have a distinct mating season, and their sperm morphology exhibits great variation according to the phase of the breeding cycle. Sperm exhibit predictable changes in their head size and length, as well as their velocity, in order to maximize a male's biological fitness when it is most relevant to do so.

Sperm morphology is also subject to immediate changes in the social environment during the breeding season. In experimentally generated 'stressful' situations, male Gouldian finches (*Erythrura gouldiae*) adjust the flagellum and mid-piece length of their sperms, which is thought to relate to an increase in stamina. This demonstrates that it's possible to strategically adjust sperm morphology in response to current conditions. Males of this species come in two basic morphologies: those with red heads, and those with black. The red-headed males are much more aggressive and dominant than the black-headed ones, and these behavioural differences are reflected in changes to sperm morphology based on the immediate social situation. The level of physiological fine-tuning exhibited by these males is astounding, and attests to the power of natural selection at multiple scales. You see, it's important to remember that evolution by natural selection can occur at both the level of the entire organism, or at any number of scales below that (organ system, organ, tissue, cellular, etc).

In the highly promiscuous fairy wren (*Malurus cyaneus*), sperm characteristics reflect the specifics of sexual promiscuity. Here, females and males are *socially* monogamous but sexually promiscuous, with distinct male:female social partnerships where both partners are likely engaging in sexual activity outside of that pair bond. Sperm traits involved in paternity offence versus paternity defence are antagonistic. Thus a fairy wren male can achieve reproductive success in two ways: first, by mating with his social partner at their nest; and second, by engaging in extra-pair copulations with other females. Males that achieve a large degree of paternity through extra-pair copulations tend to have sperm with short flagella and large heads, whereas males that are associated with higher within-pair paternity have sperm with a longer flagellum and smaller heads. It follows that these somewhat contradictory patterns of paternity should be reflected in sperm morphology, because they allow for high reproductive success for males with contrasting strategies. It's a difficult concept to wrap one's head around. You can't help wondering what came first: the social strategy or the sperm morphology? Since these factors are not mutually exclusive, there can be no definitive answer.

The sheer diversity of sperm characteristics is a sound reminder that the process of evolution by natural selection is alive and well at the cellular level. Even within one male of any species, there is tremendous variability of sperm, which is reflected in offspring produced after a major competition that few of us have ever thought much about.

The last laugh: cryptic female choice

Much of this book is preoccupied with the multitude of male strategies for manipulating, overpowering, sneaking and otherwise taking advantage of their female lovers. This section is dedicated to the tricky process known as 'cryptic female choice' (CFC), which is essentially the female's way of exercising some post-coital power over whose sperm will fertilize her eggs.

There are many reasons why a female would want to retain some kind of control over the successful sire of her offspring. In a perfect world (like the romantic scenarios envisioned by *Homo sapiens*), CFC would be completely unnecessary because monogamous couples would engage in exclusive sexual activities with each other and there would be no reason to require any kind of sperm regulatory body. Unfortunately such romantic stories are far from the norm in the animal kingdom, ditto sexual monogamy. Manipulation, deception and coercion are the rule when it comes to the contrasting biological needs of males and females, and females have evolved a number of strategies to deal with the often brutal onslaught of male-dominated sex that they encounter. Biologists suspect that CFC is widespread and important in multiple animal taxa; however, it is notoriously difficult to test because its effects need to be teased out from those of sperm competition or other physiological processes that would be interrupted by experimental intervention. Evidence from specific systems is starting to accumulate, and although the molecular mechanisms for CFC are almost completely unknown, its existence is most certainly prevalent.

In some of the more simplistic cases, females prove unable to discern a good mate from a bad one during pre-coital activities. In many insects it's extremely difficult to tell a male from a female, much less distinguish between a high-quality male partner and a low-quality one. There isn't a whole lot of 'getting-to-know-you' time prior to the copulatory event. In the insect world there are no long walks and coffee dates. Earwigs (insects in the order Dermaptera) usually mate in small dark crevasses that make it difficult to discern the quality of a mating partner prior to commencing copulation. In addition, the male's forceps, his main weapon for competing with other males, are engaged during sex and are therefore unavailable for use. This means that smaller, weaker males have the ability to interrupt the coital activities of larger males. Indeed, the rate of interruption of earwig sex is quite high, and females may have a difficult time discerning which male is the highest quality based solely on the outcomes of sex interruption and male combat. In this case post-copulatory processes become important for the female in order to maximize her biological fitness.

In red flour beetles (*Tribolium castaneum*), where both males and females mate promiscuously and frequently, the extent of CFC has been more clearly established. Males deposit their sperm-package (spermatophore) on the female's body and females control its uptake and storage. Females possess a complex sperm-storage structure that allows them to cache sperm for several months. As with many invertebrate species, in flour beetles there is a distinctive 'last male' advantage. This means that the last male to make a deposit to her sperm bank tends to be the one with the highest paternity of the resulting offspring. Sperm can be stratified, or stacked in the female's spermatheca such that the last sperm in is the first sperm out (hence, the last male advantage). However, females can mitigate this by consistently re-mating. This will allow her to expel the sperm of previous (lower-quality) suitors and refill with sperm of a higher-quality male. Overall, it makes sense for a female to initially mate indiscriminately – just to ensure that all her eggs will be fertilized. From here, she should select suitors of higher quality to create offspring with the best possible genetic makeup. When females are experimentally mated to low-quality males, the resulting broods have a mixed genetic paternity. This indicates that the females utilize some of their stored sperm to compensate for the fact that their mating partners were substandard. When females are mated with a high-quality male, the resulting broods are almost exclusively his genetic offspring. This indicates an intricate and astonishing control of females on sperm selection.

Female pygmy squid (*Idiosepius paradoxus*) do something similar, although they don't store sperm in the same way that the beetles do. Male squid attach their spermatophores to the outside of their partner's body, and multiple males can make such deposits. This leaves females with the choice of which spermatophore to accept, and they show a distinct preference for the sperm packets that come from small males. This is surprising, given what we know about large, dominant males and females' general inclination towards them. But small-statured males are much faster in their spermatophore deposition than large ones (a large spermatophore from a large male is going to take longer to extrude and deposit) and it is thought that females prefer the small ones to decrease their risk of predation during deposition. During sperm transfer, males and females are effectively affixed together, making them twice as noticeable to fish-predators (not to mention being in an awkward copulatory state and unable to react quickly). It therefore follows that females would be more apt to accept a quick sperm gift than a slow one, for both her immediate survival and the survival of her potential offspring (her sons will be more likely to be small, and preferred). What do females do with the unwanted spermatophores? Usually they eat them. That's one high-protein snack!

The ability of females in many animal taxa to store sperm is a contributing factor to their level of control over which sperm to select when the time comes to fertilize an egg. A female with the capability to keep a steady supply handy will be covered in the event a high-quality suitor is unavailable when she ovulates. Females have complicated structures that keep sperm viable from a few days up to several months, which is clearly advantageous when it comes to cryptic female choice. In some cases, having such a complex structure for sperm storage means that the 'last male' advantage is no longer relevant. In domestic chickens (*Gallus gallus domesticus*) and turkeys (*Meleagris gallopavo*), sperm from successive copulations is stored in completely separate tubules within the female's body. In addition, wild female fowl can differentially expel the ejaculates of undesirable suitors. Copulation in these organisms tends to be violent and coercive – females rarely have a choice about when and with whom copulation will take place. So it follows that we see a distinct set of well-developed CFC strategies including differential storage and expulsion of sperm.

Female ducks (family Anatidae) also have a complex genital system, with blind pouches and counter-clockwise spirals that give them a modicum of control over the extent of intromission (insertion) of a male's penis during sex. These structures have co-evolved with the spiral-shaped phalluses of male ducks, and while they cannot deter the coercive sex that is so universal in these organisms, they allow the female to limit access to her ovaries when she so desires.

There are a variety of reasons for females to select certain male suitors over others. Genetic compatibility is one, as many organisms studied in this regard have low levels of dispersal and could easily copulate with a relative. How do chickens, for example, know which chickens are kin and which are not? More often than not, this distinction is *not* made at the behavioural level. Complicated aspects of post-copulatory physiology are responsible for inbreeding avoidance. For example, female jungle fowl (*Gallus* species) bias sperm use in favour of unrelated males. Genetic differences in the major histocompatibility complex (MHC), which has control over the immune system, can be used as a proxy by which females bias sperm use. Loss of genetic diversity in the MHC could be extremely detrimental because individuals would experience suppressed immune function and have a decreased ability to combat pathogens. When copulating pairs are dissimilar in their MHC genes, a greater number of sperm successfully reaches the female's eggs, which is an astounding feat of physiology . . . but it gets even better. Females exhibit a clear bias of accepting sperm from *unrelated* males; however, this effect disappears when the females are artificially inseminated. This means that some aspect of the male's phenotype has influence on the female's physiology such that more favourable sperm is allowed easier passage to her eggs.

Another stunning example of CFC comes from the mammalian world. Female house mice (*Mus musculus*) adjust the 'fertilizability' of their ova depending on the social conditions in which they were reared. Females are thus able to physiologically alter the external properties of their eggs to make them more or less fertilizable by sperm. Let's stop and think about that for a few minutes. First, why would a female want to alter the fertilizability of her ova? Ovary defensiveness is one mechanism by which females can inhibit polyspermy, or the incidence of more than one sperm molecule entering the egg. Polyspermy is deadly to an embryo, so it is certainly something that a female would aim to avoid. When female house mice have evolved in social situations that are polyandrous, their eggs become more resistant to penetration by sperm. When females have evolved in monogamous situations, their eggs are less resistant. This type of anticipatory phenotypic plasticity has long been demonstrated in males – who can alter the sperm content and components of their ejaculate based on the scenario at hand. In this case, a polyandrous social situation will mean that males have aggressive ejaculates with high sperm counts. Monogamous males will have the opposite.

This is groundbreaking work: physiological changes to a female's eggs based on the social environment has scarcely been documented before. Having a more resistant egg when there is more likely to be a high level of sperm competition within her reproductive tract allows for successful fertilization by only the most competent sperm, and this is clearly advantageous for the female. These results were obtained through complex and controlled laboratory experiments on mice, but there are a few cases where the mechanisms of CFC have been examined in a natural setting, which is only possible for organisms that reproduce via broadcast spawning. In broadcast spawners, it's entirely up to male and female gametes to both find appropriate partners *and* to determine their level of quality respectively.

Many organisms use chemical cues to find a mate, and this happens at the cellular level between eggs and sperm. Chemical cues from eggs are used to attract sperm – but it's much more complicated than that. There are also chemically moderated gamete preferences that allow for fertilizations between genetically compatible individuals. Blue mussel sperm behaves in a fine-tuned manner such that it responds differentially to the chemical signals emitted from eggs of different individuals. For these and many broadcast-spawning invertebrates, individuals are hermaphroditic, meaning that they have both male and female genitalia. Therefore, it makes sense for physiological mechanisms to be in place to prevent individuals from fertilizing themselves, otherwise they would risk inbreeding and the increased genetic difficulties that come along with it. In addition to sperm being able to discern between the eggs of different donors, they are able to direct themselves to fertilize those eggs that will confer the highest measurable fitness benefits to the resultant offspring (i.e. non-relative or self) through chemoattractant cues that they emit. Chemical cues in the eggs contain compounds that result in sperm chemotaxis, which allows sperm to find them, and sperm chemokinesis, which increases the swimming speed of sperm, regardless of direction. It's an astounding level of chemical complexity compared to the sperm of the romantic, monogamous human.

Externally fertilizing fish provide several more fascinating examples of CFC at work. In these systems, females lay their eggs in nests on the floor of their aquatic habitat, and males fertilize them externally. Congeneric (two separate species belonging to the same genus) Atlantic salmon and Atlantic trout (*Salmo salar* and *S. trutta* respectively) have overlapping habitats throughout most of their natural ranges. These two fish species *can* successfully interbreed, although hybrids have an abnormal chromosomal makeup. At the gamete level, eggs and sperm between the two species are completely inter-fertile. Although a certain degree of hybridization is inevitable, a fifteen-day difference in peak spawning times helps to prevent it reaching high levels. If eggs of either species are put into contact with sperm of the other (in an experimental scenario), successful fertilization is guaranteed. However, if eggs of either species are placed into contact with sperm of both species, the conspecific sperm will always win fertilization precedence, suggesting that some aspect of CFC is in place to favour the sperm of same-species males. Ovarian fluid is a viscous, protein-rich gel that is secreted by females along with their eggs, and chemical factors contained within it are responsible for increasing sperm lifespan and swimming ability.

Another compelling aspect of salmonid ovarian fluid is observed between farmed and wild individuals. Fish farming is a major industry in many parts of the world – and escapements (breakouts) from open-ocean fishnets are a general problem. Without getting into a discussion about the logistics of fish farming, it's worth noting that the ovulatory fluid of farmed females may have detrimental effects on sperm of wild males. Atlantic cod (*Gadus morhua*) farmed off the east coast of Canada frequently escape from their nets and fraternize with wild populations. Wild males actively court farmed females that have escaped (sounds like a Disney movie gone wrong), but their fertilization success is low due to the fact that components of the farmed female's ovulatory fluid negatively affect sperm motility and ability to fertilize. It's thought that the adverse effects of the ovulary fluid are due to nutritional deficiencies in farmed females. Regardless of the cause, it provides a striking example of the power of ovulatory fluid on sperm dynamics.

When a boy is a girl too

As members of a species with separate sexes, humans tend to forget about the suite of organisms out there that simultaneously carry male and female genitalia. Hermaphroditism is present in twenty-four of the thirty-four animal phyla (taxonomic groups), and is quite common in the invertebrate world. In fact, for many organisms hermaphroditism is the rule rather than the exception. When animals are both male and female at the same time, the scene is set for a good deal of biological conflict. We have already seen many examples of conflict between the sexual needs of males and females, so what is an organism to do when such conflict exists within itself? Hermaphrodites from various phyla display behaviours and physiologies consistent with the notion that sexual strife is alive and well at the level of the single organism. Aptly termed the 'hermaphrodites' dilemma', each individual has the urge to be sexually gregarious, courtesy of its male self, and sexually choosy, thanks to its female self. This dilemma has led to the evolution of some extremely bizarre and violent sexual rituals.

Let's envisage two hermaphrodites meeting. There are at least three scenarios to consider. First, there are reciprocal copulations where both individuals act as both partners during an encounter. Two individuals, two penises and two vaginas. Second, there is unilateral copulation, where one individual acts as a male and the other acts as a female. Two individuals, one active penis and one active vagina. Kind of like traditional sex, except that the male partner also has an inactive vagina, and the female partner has an inactive penis. Imaginably, it is more biologically advantageous to be the male in said scenarios because it is much cheaper to transfer gametes and move on to the next partner than it is to be the one who is left to maintain the fertilized embryos.

Lastly, although not all species are capable of it, many hermaphrodites can self-fertilize. One individual, one penis and one vagina. Biologists often argue that self-fertilization should maintain a baseline of the sexual activities of any individual as a safeguard for fertilization assurance. Once an individual has transferred enough sperm to ensure total fertilization, said individual should invest energy in finding higher-quality sperm (as a safeguard against inbreeding depression), and transferring sperm of their own. So within any one individual there is both auto-sperm (self sperm) and allosperm (sperm donated by others). Interestingly, only a tiny portion of the sperm received by an individual (as low as 0.025–0.1 per cent) is stored for further use in copulations. The lucky sperm that manage to reach storage can remain in the spermatheca from a few weeks to over a year, depending on the species. Most of the sperm that an individual hermaphrodite receives is digested in an organ called the bursa copulatrix – clearly a function that has evolved on behalf of the female part of the hermaphrodite. Sex and sperm transfer do not necessarily equal reproductive success in hermaphrodites. These are some seriously complex ideas to wrap one's head around, so they are perhaps best illustrated by examples.

One of the more studied groups when it comes to sexual

strategies of hermaphrodites is the Gastropoda (slugs and snails). We begin our discussion with nudibranchs or sea slugs, where hermaphrodites exhibit efficient sperm digestion and traumatic hypodermic stabbing. For example, in the sea slug *Siphopteran quadrispinosum* 'males' use a syringe-like penile structure to pierce the integument of their sexual partners. This traumatic injection carries a prostate fluid that is thought to contain allohormones that change the sexual behaviour of the receiver. Subsequent to the initial injection, the male anchors his penis to the female's genital opening using four or five large hook-shaped spines at its base. In addition, the tip of the penis carries up to thirty small spines, so the entire act of sex in these organisms can cause considerable wounding to the receiving partner. Both reciprocal and unilateral matings take place in this species, and the 'courtship' ritual involves a long and strenuous struggle. There is a small window wherein a 'stabbed' partner can still carry out a jab of its own, but if this individual doesn't manage to get in an immediate reciprocal blow 'she' will be forced into exclusively playing the female role. The first individual to successfully stab the other will engage in a number of behaviours to minimize his chances of getting reciprocally stabbed, including using his parapods (arm-like appendages) to swat away the partner's penis, rapid circling and minimal exposure of their ventral side (where the stabbing takes place). Clearly, it's a lousy deal for the female partner. In addition to the wound and risk of infection from the stabbing, 'she' is forced to play the more expensive sexual role (see below).

Although one might assume that 'females' would want to mate as little as possible, field observations show that *S. quadrispinosum* individuals copulate in the female role once or twice per day. This is much higher than would be required for basic fertility, so what could possibly explain this behaviour? It's thought that the increased copulations bring indirect (genetic) benefits as well as direct nutrition benefits of the ejaculates. Fecundity levels of 'females' have been demonstrated to max out at the mid-level, and drop at both low and high levels of sexual activity due to inbreeding and trauma respectively.

Five separate *Siphopteran* species demonstrate that there is distinct, repeatable variation with respect to *where* on the body the stabbing occurs. Although with *S. quadrispinosum* the stabbing happens in the anterior foot region, the wounding in another (as yet unnamed) species occurs with 100 per cent accuracy on the forehead. It's thought that both the location and the remarkable precision with which the injections occur mean that allohormones deposited on to the cerebral ganglia could have an immediate and substantial effect on the post-copulatory behaviour of stabbed individuals via manipulation through direct contact with the central nervous system.

Another bizarre mating ritual is demonstrated by the sea slug *Chromodoris reticulata*. To date, this is the first species observed to have a completely disposable penis. Subsequent to each and every copulatory event, the 'male' individual autotomizes (casts off) his penis. It is then regenerated and the lucky males are able to successfully copulate again within twenty-four hours. It turns out that there is complex morphology behind the ability of males to cast off their members after each mating. The entire penile structure is quite large and attains a spiral shape within the male's body. Its overall length is enough for three copulations, which is why the male can regenerate it so quickly after autotomizing the previous one. Essentially, once a copulation has taken place and the end of the penis is autotomized, the next part of the penis is uncoiled from the spiral and makes its way out of compression ready to inflate and copulate. Think about toilet paper coming off the roll. The autotomized members play a role in removing sperm from previous reproductive partners, as observations show them to have numerous backward-pointed spines and to be covered in masses of sperm.

From sea slugs we move on to land slugs, which have bizarre mating rituals of their own. The copulatory process takes between seven and twenty hours, and it is both graceful and horrific. Something called a phally polymorphism exists in several species of the genus *Ariolimax*, whereby there are a variety of penile forms within individuals of the same population. Some have a complete functional penis (euphallic), whereas others completely lack a penis (aphallic). Still others have a reduced, non-functional penis (hemiphallic). It's thought that the truncated (hemiphallic) versions of the penis are the result of a delightful process called apophallation – where individuals chew off the penises of their partners. Yes, penis loss is directly attributable to the mating partner, and no, it cannot be regrown. The exact reasons for the existence of apophallation are unknown, although it is hypothesized that the lengthy members of mating partners get so entangled with one another that two individuals cannot get themselves apart otherwise.

While apophallation explains the existence of the hemiphallic state, it does not explain the existence of the aphallic state – where the penis (or any related penile musculature) is completely absent. Despite not having a penis, aphallic individuals still have functional ovotestes (the organs that create sperm and egg), meaning that the degree of gamete production is the same in euphallic and aphallic individuals. These individuals can use either their own sperm or the sperm from a sexual partner to create offspring, but it's physically impossible for them to give out sperm to other individuals. Why would such a strategy evolve? To be honest, the details on that are not yet clear. The evolution of sexual strategies in hermaphrodites continues to perplex even the top experts in the field.

We will remain in the class Gastropoda for this next example, this time focusing on those with a shell. Here I discuss the infamous love darts of land snails. What exactly is a love dart? These are either calcareous or chitinous structures that are used to stab would-be sexual partners prior to copulation. They come in diverse forms in most land snail species, ranging from simple cone-shaped darts to massive structures with blades and teeth. All known love darts come with a dose of chemical manipulation in the form of a glandular product that causes conformational changes in the female reproductive system. Its primary function is to halt sperm digestion in the bursa copulatrix (the organ specifically designed to digest excess sperm). Hormonal changes caused by the glandular secretion prevent the digestion of the shooter sperm, which is significantly beneficial both in terms of an increased number of sperm maintaining their integrity and also a degree of refractory behaviour experienced by the dart's receiver (due to being 'full' of sperm). Indeed, in the snail *Euhadra quaesita* the remating interval of individuals is greatly increased when they have been stabbed with a love dart versus not (fifteen days versus seven), which clearly demonstrates the refractory implications of being punctured by Cupid's arrow. In the snail *Lymnaea stagnalis*, additional compounds carried with the dart manipulate and decrease the maleness of the individual receiving the dart, which thereby increases the receiver's energy into female activities. This is excellent news for the dart's shooter, seeing as it's his sperm that are now within her reproductive tract.

Overall, the presence of a love dart seems to increase the reproductive success of the male part of a hermaphroditic individual, a theme we saw in the sea slug example above. In fact, several hermaphrodite 'males' transfer ejaculate components to their partners that serve to decrease the male function in the receiving partner. For example, biologists have been able to isolate two specific seminal fluid peptides (LYAcp5 and LyAcp8b) from ejaculates of the hermaphroditic pond snail (*Lymnaea stagnalis*). These peptides have a substantially *negative* impact on the male functionality of their partners. It's possible this strategy evolved because it is advantageous for the delivering partner to have the receiving partner spend more time on the female role – especially if processing the sperm of the delivering partner. After all, playing the female role is more costly in terms of energy expenditure on eggs, and so it could easily be preferable for a hermaphroditic 'male' to have the other partner be the girl. In addition, a 'male' can enhance his own reproductive success by limiting the male function in the receiver. Think about it from his perspective: if my partner makes a lot of sperm and then attempts to deliver that sperm to other partners, this sperm is then potentially competing with my sperm in those partners. Moreover, increased sperm delivery in my partner means that my partner may receive more sperm in return, which will then compete with my sperm to fertilize the eggs that I am hoping to solely fertilize. The reproductive ecology of hermaphrodites is in a realm of its own, because the complexities that arise when one is both male and female at the same time mean that organisms can essentially be at war with themselves.

We will now depart from the world of molluscs and head over to another invertebrate phylum, the Annelids, that also display the remarkable products of sexual coevolution in hermaphrodites. Good old earthworms (*Lumbricus terrestris*) use a set of 40–44 copulatory spines or setae that pierce their partner's skin during copulation (see 'Fifty Shades of BDSM in the Animal Kingdom'). These setae are located on the ventral side of the body on segments 10, 26 and 30–38, with two pairs per segment. Earthworms have four spermathecae (sperm storage organs) that are located in the region where most of the copulatory setae are pierced. These organisms are simultaneous copulators, and glandular products transferred by the copulatory setae are thought to function in increasing sperm uptake as well as in maintaining an even distribution of sperm within the four storage organs of the partner.

Fascinating as they are, tales of hermaphroditic copulation in earthworms pale in comparison to the sexual exploits of their close relatives, the flatworms (phylum Platyhelminthes). The 'penis-fencing' rituals of flatworms from diverse species describe a courtship routine whereby individuals 'rear up' and attempt to be the first one to stab the other with their razor-sharp penile members. These are fairly violent interactions, and they take much longer when both individuals have their penises drawn and ready (picture an old Western movie where two cowboy heroes have their guns drawn – except in this case it's erect penises). Most inseminations in flatworms are unilateral, meaning that the first successful stabber gets to play the male role. Subsequent to receiving sperm, the female-acting member of the sexual encounter almost always bends down on to itself, places its pharynx over its own genital opening, and proceeds to engage in a kind of sucking behaviour, which can only be interpreted as a mechanism of post-copulatory cryptic female choice. There are two major themes to the mating behaviours exhibited by many flatworm species in the *Macrostomum* genus: first, the reciprocal mating syndrome, and second, the hypodermic mating syndrome. Both have contrasting sexual characteristics that are dependent on specific aspects of a species' mating behaviour.

The *reciprocal mating syndrome* describes a situation in which we expect female resistance traits (such as sperm digestion and post-copulatory sucking) to evolve, as well as male persistence traits that defend against such female control. In species that exhibit this kind of scenario, we see extremely complex sperm morphologies that allow for its success. Sperm are deposited into a female's genital opening, and although they lack flagella (the tail structure common to most sperm), they are highly motile and have a front feeler, which allows them to become anchored within her. Sperm also have two lateral bristles that may serve an anchoring function (especially in light of the post-copulatory sucking that takes place). Females in the reciprocal mating syndrome category exhibit a thickening of their antrum – the area on her reproductive tract where the majority of sperm lands.

When this kind of coevolutionary violence between the maleness and femaleness reaches a 'point of no return', we see an evolution towards forced unilateral sperm donation and hypodermic injection rather than deposition within her genital opening. The *hypodermic mating syndrome* describes such a scenario, where sperm are stabbed into the female's body cavity. No post-copulatory sucking behaviour exists in this scenario (why would it when the sperm is directly injected?). Sperm remain aflagellate, but they no longer have the front feeler or the lateral spines. Females do not exhibit a thickening of the antrum, which makes sense because sperm are no longer deposited there.

These two strategies explain the majority of mating behaviours in flatworms, the reciprocal mating syndrome being driven by sexual conflict and the hypodermic mating syndrome being driven by sexual selection to bypass traits for female resistance. Hermaphrodites are complicated.

There is vast diversity within the hermaphroditic sexual behaviours of any population depending on other ecological aspects including sexual status. For example, the flatworm *Macrostomum lignano* engages in both simultaneous and unilateral copulations. Virgin pairs tend to copulate more than do previously mated individuals, suggesting a higher willingness to donate and receive sperm when there hasn't been any previous exchange. From the perspective of fertilization assurance, this makes sense; and it follows that the engagement in post-copulatory sucking behaviour also hinges on whether an individual has previously mated or not. Virgin individuals have a lot of ejaculate to give out, and they also require some sperm if they are to avoid the need to self-fertilize. Species from many different animal phyla have been shown to engage in sophisticated and minute changes to the allocation of maleness or femaleness depending on the ecological scenario at hand. The biology of hermaphrodites is as fascinating as it is amazing – and almost impossibly complex.

Sex thy self

Masturbation – 'self-manipulation of one's genitalia' – is a prevalent activity in many organisms, and humans are no exception. Masturbating is a pleasurable way to release stress, it's easy, doesn't require much room, and it's free of charge. Moreover it's a biologically relevant practice that is completely natural and necessary for healthy sexual function. From invertebrates to vertebrates, mammals, cetaceans and of course primates, self-pleasuring is rampant and important. The 'sexual outlet' hypothesis essentially states that in absence of having a partner for copulation it makes sense to self-satisfy. Sexual arousal and tension can be effectively appeased in this way, allowing individuals to proceed with other biologically relevant activities (feeding/grooming/avoiding predators) once the deed is done. We can reasonably assume that in species of mammals where orgasms exist, the sexual outlet hypothesis represents a satisfactory explanation for the existence of self-pleasuring.

Primates are persistent masturbators. Diverse groups including chimpanzees, bonobos, baboons, orangutans and macaques utilize a complex suite of masturbation techniques ranging from simple genital stimulation with a hand to the use of twigs, leaves, blades of grass or other inanimate objects. Both males and females engage in masturbation; females have been observed to stimulate their external genitals (clitoris) with their fingers or any number of objects, as well as to insert implements directly into their vaginas. Male primates most often utilize their hands for manual stimulation, although they have been observed using other objects or surfaces. In one display of sexual ingenuity, a male orangutan (*Pongo* species) created his own 'sex toy' using a large leaf, through which he poked a hole with his finger. He then proceeded to thrust his erect penis through the hole for additional stimulation. In another, a male chimp (*Pan troglodytes*) at a zoo in Honolulu used a frog as a sex toy, killing it by repeatedly thrusting it over his erect penis. Both examples clearly fit into the context of the need to relieve sexual tension.

The 'fresh ejaculate' hypothesis explains the existence of masturbation in the context of male physiology. It's important for boys of any species to keep their ejaculate contents in a fresh state. Sperm can decline in both motility and quality with age, which is bad news for the males that possess it and for the females that receive it. The best thing that a male can do to keep his sperms in tip-top condition is to masturbate and discard those that have experienced the ill-effects of ageing. For example, male house crickets (*Acheta domesticus*) routinely expel sperm-packets outside of their sexual activities with females. Females have the ability to store sperm delivered by males during copulation, and young sperm have a competitive edge against older sperm in the female's sperm storage organ. In addition to young sperm being competitively dominant, females are also most likely to keep or utilize the freshest and highest-quality seeds to fertilize their eggs. Indeed, experimental dissections of sexually mature females reveal that most sperm in her storage organ is young and fresh (less than six days old), meaning that males can improve their reproductive success by providing a female with a package of young sperm. Autonomous sperm expulsion is the official term given to cricket masturbation, in which males utilize their ability to expel spermatophores outside of copulation.

There are other well-substantiated hypotheses to explain the existence of self-pleasuring, such as the 'sexual arousal' hypothesis, coined to describe the incidence of masturbation in a free-ranging population of *Rhesus macaques*. These primates have a complex social system wherein males have various levels of social status. Those of high rank are in control of much of the sexual activities of the group, and tend to receive the bulk of the action. As with all primate species, male macaques engage in a substantial level of masturbation, although low-ranking males generally indulge in the practice more than high-ranking ones. This isn't terribly surprising, given that the high-ranking males have a lot of sex with females and so they need not spend as much time on masturbating. However, the low-ranking masturbators experience a very low rate of ejaculation – only 15 per cent. So the self-pleasuring behaviour of low-ranking male macaques cannot be explained by either the sexual tension hypothesis or the fresh ejaculate hypothesis, both of which require emission to occur. According to the sexual arousal hypothesis, low-ranking males keep themselves in a state of constant sexual arousal so that they can be 'ready' to mount a female on short notice if the opportunity arises. Since these males are generally not permitted to have sex by their high-ranking competitors, it makes sense for them to do what they can to minimize the time it would take them to ejaculate into a female.

A similar scenario occurs in male marine iguanas (*Amblyrhynchus cristatus*) of the Galapagos archipelago. There are substantial size differences between sexually mature males. The largest are the most aggressive, and these guys hold territories in which females copulate, nest and feed. To gain access to a female, smaller males must sneak on to the larger males' territories and copulate while the dominant males are otherwise occupied. In general, it takes approximately three minutes of intercourse for a male iguana to ejaculate. The large, dominant males have no problem achieving success with the females, so we can imagine that those are three minutes well spent. However, for the small males three minutes is prohibitively long. Prior to ejaculation it is common for small males to be forcefully separated from their chosen females (by the dominant male) and removed from the territory. So, in order to increase their reproductive success they masturbate before having sex and store the prepared ejaculate in small pouches near the penis. The ejaculate can then be thrust into a female immediately upon mounting, and this strategy serves to increase the reproductive success of low-ranking males by a respectable 41 per cent. The sexual arousal hypothesis may explain the existence of masturbation in a wide range of species where a size-dominance hierarchical system exists.

Masturbation has been observed to occur in African Cape ground squirrels (*Xerus inauris*) in a post-copulatory context. These are highly promiscuous rodents where populations can have a skewed operational sex ratio. Males vastly outnumber females, and a female may mate with up to ten males during her three-hour oestrus (fertile) phase. A few aspects of the self-pleasuring rituals of the squirrels are unique and speak to an entirely new hypothesis for their existence. First, high-ranking males are more likely to be the ones masturbating. These males are also the ones seeing most of the sexual action, which rules out the sexual tension hypothesis. As I mentioned above, the masturbation activity takes place *after* copulation, which rules out the fresh ejaculate hypothesis. If a male has already provided his sperm to a female, what's the point in ejaculating again, after the fact? It seems that the answer lies in a self-grooming context. Since there is such a high level of promiscuity in these squirrels in a very short space of time, biologists hypothesize that individuals masturbate as a way to clean their genitals and prevent the spread of sexually transmitted infections. Both males and females engage in the self-pleasuring behaviour; for males it is described as an oral and a manual ordeal. Males take their own penises into their mouths while using their forepaws for additional stimulation. They thrust their torsos back and forth until ejaculation occurs, at which point the ejaculate is normally consumed. A similar post-coital cleansing of the genitals has been observed in Gunnison's prairie dogs (*Cynomys gunnisoni*), where males and females emerging from underground trysts engage in oral self-sex. Experiments on genital grooming in rats show that if males are prevented from oral self-cleansing of their genitals subsequent to copulation, they have a significantly increased chance of contracting microorganisms present in the female's vagina. There is evidence to suggest that there are antibacterial properties in the saliva of the rats, prairie dogs and squirrels that could lend themselves to fighting infections.

I'll end this section with an interesting, yet horrible example of human attitudes towards sex. It wasn't too long ago that masturbation was regarded in Western society as a sinful act, something dirty and impure. And not just in human beings. Despite the fact that masturbation is, as we have seen, commonplace in the animal kingdom for logical, well-defined reasons, humans frowned upon displays of such 'lewd' behaviour. In male equids – an animal group that includes horses, zebras and donkeys – spontaneous erections followed by penile movements (rhythmic bouncing and pressing against the underside of the abdomen) are extremely common. In fact, during waking hours this kind of 'masturbation' occurs approximately every ninety minutes. To prevent such an unsightly thing occuring during a competitive show, the horse-training fraternity adopted a range of 'anti-masturbatory' devices such as 'stallion rings': constricting bands of metal or plastic that bind the penis in an attempt to prevent erections. Other measures include attaching nail patches or abrasive brushes to the underbellies of horses in the place where an erect penis would make contact; fixing plastic or metal baskets to cover the entire genital region; or even making the stallions wear collars that deliver electric shocks – all in the name of preventing a perfectly natural process. Ironically, in many cases these techniques result in erections occurring with greater frequency – as though the natural physiology of the horse is attempting to win the battle over its unnatural enslavement. Sexual dysfunction is a common clinical problem with prize-winning stallions; in a series of experiments designed to test the effects of these measures on equid libido, it was found that subsequent to 'anti-masturbatory' treatment, males showed a marked decrease in normal sexual function towards receptive females, and these results persisted for a number of months. The inability of members of our own species to recognize the basic biology of their most prized animals is something that makes me very sad indeed.

The big O

Pleasure isn't something that's come up in this book too often. The unfortunate truth of the situation when it comes to pleasure is that it isn't all that commonly associated with sex in the animal kingdom. There are several reasons why this is the case. First, we cannot simply ask animals how something felt. Was it good for you? Did you come? It's utterly impossible for us to know. Of course in species where torturous sexual techniques involve participants getting stabbed, strangled or killed, it seems safe to assume that this would *not* afford physical pleasure. Likewise there are some species for whom sex happens in a way that does appear to offer a favourable sensation. Case in point: what I lovingly term the 'Magnum P.I.' hypothesis about sexual activities in Mexican guppies (*Poecilia sphenops*).

Epidermal outgrowths (aka moustaches) on the upper maxilla of some males appear to provide them with an advantage when it comes to sexual reproduction. Females prefer to associate and copulate with moustached males despite there being no clear biological or physiological advantage to the extra bristles. The inevitable conclusion is that the filaments provide pleasurable stimulation prior to copulation, since males perform a 'nipping' routine that involves repeated contact of their moustaches with the female's genital area. Another example of what would seem to be pleasurable stimulation comes from the invertebrate world. Earwig (*Euborellia brunneri*) males exhibit phenotypic variability in their genitalia. To put it bluntly, some males have extremely long penile appendages, and others not so much. Males with large members are otherwise no different from those with short ones. They aren't more socially dominant, larger or healthier; the difference in their genital morphology results from a simple genetic characteristic. One might conclude, based on what we know about female choosiness, that short-membered males would experience a greater level of sexual success based on the time it takes for sperm to move through a long tube versus a short one. A female would be better off to copulate quickly in order to save time for other biologically relevant activities. However, just the opposite is observed in the wild! Females repeatedly choose to copulate with long-membered males over short ones. Why? According to researchers (and I quote): 'During copulation, the genitalia of certain males may elicit more favourable female responses through superior mechanical or stimulatory interaction with the female reproductive tract.'

Now, it's utterly impossible for us to decisively conclude that sexual pleasure is at the root of fish moustaches and extra-long earwig penises. However, there are other examples of structures where sexual satisfaction can be more directly inferred. Buffalo weavers (*Bubalornis* species) are a group of birds that are known for their unique genital structures. Males possess an enlarged phalloid organ located anterior to their cloacas. Females have a vestigial version that is non-functioning. The phalloid is non-erectile and has no sperm duct; in fact, it is entirely comprised of connective tissue and is not homologous with any other penis or penile structure in the animal kingdom. Its mere existence has puzzled scientists for hundreds of years. Recent work has demonstrated that the phalloid plays a key role in facilitating ejaculation in males, and it's hypothesized that buffalo weavers are the only male birds to experience orgasm. Interestingly, the phalloid is assumed to be stimulatory to the female as well. Copulation in buffalo weavers is a prolonged experience, which is thought to be a male strategy for safeguarding paternity once ejaculation has occurred. How's a male going to keep a female happy subsequent to sperm transfer? Apparently by continuing to stimulate her with his phalloid organ. It seems buffalo weavers could be the only birds in the animal kingdom to realize the luxury of sexual pleasure!

For mammals, the situation *should* be a little different. All mammalian females have a clitoris, which in our own species affords a great deal of sexual pleasure, including orgasm. Does it serve the same purpose in other mammals? It's difficult to say, although some general conclusions can be drawn with respect to the kinds of physiological processes that occur when human females experience climax. For primates especially, many females exhibit an increase in heart rate, respiration and blood pressure associated with vaginal and clitoral stimulation. Involuntary, rhythmic muscle contractions in the vagina, uterus and anal sphincters occur at orgasm in human females, and these same contraction patterns are demonstrated in other primate females, leading to the obvious conclusion that other ladies experience orgasm as well. In chimpanzees, bonobos and macaques, female orgasm is also associated with a reach-back response, where females will clutch their sexual partner and reach backwards while emitting distinctive vocalizations and exhibiting a characteristic facial expression. Overall it seems abundantly clear that happy primate ladies all over the world are experiencing orgasm. The more interesting debate in the scientific sphere concerns *why* such a phenomenon occurs.

There are two main hypotheses for the existence of female orgasm. The first is called the byproduct hypothesis, and it states that the female orgasm is a mere side effect of its occurrence in males. Basically, since the structure and function of the penis is so critical to male reproductive success, and the clitoris is a vestigial penis, females enjoy the benefit of orgasm for no real evolutionary reason. The second major hypothesis for the existence of female orgasm is called the mate choice hypothesis, and it states that the orgasm evolved to function in mate selection, to increase the probability of successful fertilization by high-quality males. The majority of evidence suggests that while there is a common evolutionary origin to the potential for male and female orgasm (i.e. the clitoris as a vestigial penis), there are vast differences between male and female orgasms that suggest that they have evolved in different directions since their origin.

Male orgasms are tactile, and can be elicited in predictable fashion. Female orgasms are more elaborate and psychologically complex, and they can happen multiple times in direct succession. While it's implicitly obvious that male orgasm works to promote reproductive success, several aspects of female orgasm suggest that it too acts to promote conception. Vaginal contractions can stimulate male ejaculation, and associated pressure changes in the uterus act to draw sperm inward (delightfully termed the 'up-suck' effect). In addition, peristaltic uterine contractions stimulated by the release of oxytocin at orgasm act to transport sperm through a female's reproductive tract. From an evolutionary point of view perhaps the most telling fact is that female orgasms have been found to occur more frequently at or near ovulation, the peak fertility period in a female's monthly reproductive cycle.

For me one of the most interesting aspects of female orgasm in humans is that its frequency is directly related to specific partners. There is an unambiguous link between pair bonding and orgasm in females, which is thought to be mediated by the hormone oxytocin. If a female is unhappy (or merely tolerant) with her sexual partner, orgasm is highly unlikely. This kind of connection between female orgasm and high-quality partners is also observed in other primates, many of which experience greater frequency of orgasm with dominant male partners. In an experimental scenario, low-ranking female Japanese macaques (*Mucaca fucata*) paired with high-ranking males exhibited the highest rates of orgasm whereas high-ranking females paired to low-ranking males exhibited the lowest. Although not particularly surprising, it seems to confirm that we human females are not the only ladies for whom sexual enjoyment is an emotional experience as well as a physical one.

Bestiality

It's the stuff of obscure pornographies, it's something you've heard tales about, it's a mix of eroticism and craziness that is almost unbearably disgusting yet curiosity-piquing all at the same time. It's bestiality. Sex between species. It actually happens a lot in the animal kingdom, and no, not between some *Homo sapiens* and a horse or dog. There are animal examples, too, such as incidents of sexually mature male fur seals (*Arctocephalus gazella*) repeatedly attempting to copulate with adult king penguins (*Aptenodytes patagonicus*) in Antarctica. One isolated incident was reported several years ago where a seal, weighing in at over 100 kilos, subdued the penguin (no larger than 20 kilos) and alternated between resting its weight on the tiny body and attempting to thrust its erect penis into the penguin's cloaca. The penguin was totally subdued after the attack but appeared to be otherwise physically unscathed. Three further acts of sexual coercion by fur seals on king penguins have recently been documented, and they all follow a similar pattern. First, the seal chases the penguin, then captures and mounts it and repeatedly attempts copulation. Several bouts of penis-into-cloaca thrusting occur with rest periods in between (although the penguin remains pinned down).

Biologists have observed intromission of the fur seal penis, and between-the-legs bleeding of penguins afterwards.

Although the exact reason for this inter-species sexual activity is unknown, it is hypothesized to be due to an overabundance of male fur seals in specific populations. Male seals that cannot find a female partner become sexually frustrated and take out their angst on unsuspecting male penguins. In all but one observed case the penguins were released afterwards; even so, it's likely there would have been an immediate decrease in their biological fitness after the rapes.

On the whole, bestiality in the animal kingdom is not as dramatic as the examples above. For many invertebrates, fish and amphibians, sex between species (called sexual interference) occurs for a variety of reasons, including a simple lack of recognition.

Many frog species exhibit something called explosive breeding, which essentially entails a short breeding season wherein there is intense male:male competition for mates. Often, many males will attempt to court the same female, since the latter are only sexually 'available' for a short window of time. In some species where adults are terrestrial but mating takes place in the water, females end up being crushed or drowned by sex-crazed males. This loss of life during explosive breeding is the unfortunate reality for females of the Amazonian frog *Rhinella proboscidea*; in fact it is such a frequent occurrence in this species that males have evolved a 'functional necrophilia' reproductive strategy, whereby they promote extraction of the oocytes from the abdomens of dead females and fertilize them in a post-mortem necrophiliac fashion. This strategy might sound appalling, but it functions to increase the reproductive success of both the females and the males.

The point is, explosive breeding is a desperate time for males of many anuran (frog and toad) species. It may be their only chance to pass on their biological blueprints – and such intense competition over a short space of time can undoubtedly lead to courtship mistakes. Such is the case in several species of frogs that exhibit both explosive breeding and amplexus behaviour, a form of mate-guarding where a male will vigorously clasp the female to prevent her from engaging in sexual activity with others. Males performing amplexus often clasp females of the wrong species, leading to misdirected courtship that prevents both male and female from achieving a high level of biological success. Male frogs have also been known to mistakenly grasp dead females, salamanders, floating debris or even the hands of a human trying to observe the action.

There are several examples in nature where misdirected court-ship can lead to more serious consequences than the decreased biological success of a few individuals. When invading species threaten the existence of native ones, reproductive interference can have dire consequences. In a subsequent section on sexual cannibalism, I explain how female invasive South African praying mantids (*Miomantis caffra*) mimic the pheromonal cues of native New Zealand aphid females (*Orthodera novaezeandiae*), luring unsuspecting native males to their deaths. However, many other examples of reproductive interference leading to native species demise have to do with aggressive (invasive) males courting native females. For example, populations of the endangered frog *Rana latastei* are imperiled by misdirected courtship behaviour from an invading frog species *Rana dalmantina*. There are a number of forces at work here, from the aggressive male heterospecific courtship behaviour of the latter species, to the inability of females of the former species to recognize their own kind. As conspecific males (in this case male *R. latastei*) become less and less abundant, females not only lose their ability to discern them from closely related frogs, they are also faced with a 'need to be fertilized' when they have an abdomen full of mature eggs. This unfortunate predicament sets the scene for rapid extinction of the native species because females are more likely to keep copulating with the invasives.

A similar scenario is playing out for a native endangered fish species (*Skiffia bilineata*) in Mexico. Invasive Trinidadian guppies (*Poecilia reticulata*) are wreaking havoc on the native fish due to the aggressive and persistent courtship by males of the latter on females of the former. In fact, the male guppies both court and attempt to force copulations on female native fish even when females of their own species are present in excess. This aggression amounts to sexual harassment and female Trinidadian guppies have evolved a mechanism to deal with it: namely by segregating themselves from males of their own species in deeper, faster water. The native fish females have not evolved the same tools to 'deal' with the invasive males (i.e. they remain in the same areas with them) so their demise via sexual harassment is practically inevitable.

Misdirected courtship also has the potential to influence the overall distribution of species within certain microhabitats. For example, the highly aggressive seed bug (*Neacoryphyus bicrucis*) has been found to limit the biodiversity and composition of other insect species on the host plants it inhabits. They do this through aggressive, heterospecific courtship of females, creating a situation of 'pseudo-competition', a phenomenon that on first observation appears as though the seed bugs are out-competing other species for a certain resource (i.e. the plant habitat). However, carefully designed exclusion experiments reveal that the true cause of the lack of biodiversity has everything to do with misdirected (highly aggressive and damaging) courtship on females of other species.

The previous examples I've discussed have largely been about attempts at heterospecific *courtship*. While such actions can be annoying, stressful and ultimately damaging to the biological fitness of both males and females (due to time lost on other biologically relevant activities while either giving or receiving heterospecific courtship), they do not generally result in direct injury or death. Misdirected courtship attempts usually end at the courtship stage and do not result in copulation. There is another entire category of reproductive interference dedicated to what happens when courting males take it up a notch and are able to successfully copulate with interspecific females. This is called heterospecific *copulation*, and on the whole such mismatched coitus does not end well.

Evolution *should* select for individuals that employ ways of discerning members of their own species from any others. Reproductive success of those engaging in sex with organisms of another species should therefore be drastically reduced. Indeed it is. The 'lock-and-key' hypothesis holds that there is only one specific key (male genitals) that fits inside a certain lock (female genitals), and while this theory isn't without its own controversies, there are many examples where it holds true. Just ask the genitally mutilated individuals that have gotten it wrong. In the case of two parapatric (some overlap in geographic niche) carabid beetle species (*Carabus maiyasanus* and *C. iwawakianus*), males copulate with females of both species indiscriminately. Ladies that are fertilized by the wrong gentlemen *can* lay these fertilized eggs, but the rate of successful fertilization is low, and the survival of the offspring is negligible. More often, females suffer fatal injuries from copulations with heterospecific males due to rupturing of their vaginal membranes. Males don't fare much better, as they often experience broken genitals following mismatched copulations and this effectively prevents them from achieving successful fertilization in the future.

In addition to the wounds sustained by females from being

stabbed with the wrong machinery, they are also subject to increased risk of infection from said wounds. A detailed study of the consequences of heterospecific copulation in two species of fruit fly (*Drosophila santomea* and *D. yakuba*) used fluorescently labelled microbeads strategically placed on the male genitalia. When sex is between species, the haemocoel (body cavity) of the females is more likely to be invaded by foreign substances. This was brilliantly demonstrated by the presence of the fluorescent beads that had been transferred via their partner's genitals. It's bad enough to have a morphologically incorrect penis stabbing through your body without being subjected to an increased rate of infection as well. Even worse, imagine having an entire dead male of the wrong species attached to your nether regions. We've seen this before in a within-species context, but it happens between species as well. This is the unfortunate circumstance for female reptile ticks (*Aponomma* species) that happen to come into contact with the wrong males in areas where their geographic distributions overlap. In these situations, it is possible for two (or even three) different species to be lodged on the same reptile host. Males are generally indiscriminate, and will undertake forced copulations with females of the wrong species. This results in a mess of genital confusion such that they cannot remove themselves from her nether regions after the act. Thus, their entire body becomes the ultimate mating plug, though sadly it is of no biological consequence.

Engaging in cross species sex is clearly bad news for both males and females, but females are most likely to suffer increased mortality as a result. This could be a major reason why male and female plant bugs (*Coridomius taravao*) have both evolved to be morphologically similar to *males* of a sympatric species (*C. tahitiensis*). These two species of plant bugs employ traumatic insemination – whereby males stab their sharp penile weapons into the female's body cavities, irrespective of her genital opening. However, females of each species are stabbed and inseminated through different parts of their abdomens. These areas are called paragenitalia, as they have become a kind of genital receptacle outside of their primary genital openings (which are usually atrophied and useless). There's little in the way of pre-copulatory courtship for either species; the male merely grabs a female and begins the mating (stabbing) ritual. Female *C. taravo* are much smaller than both male and female *C. tahitiensis*, which subjects them to a greater degree of harassment from males of the latter. In order to combat this, females of the smaller species have evolved to actually appear similar to males of the larger one! Sexual mimicry to avoid unwelcome harassment is rather common in the animal kingdom, but this is the first documented case of mimicry that has evolved in order to deter harassment between species.

In two species of bean weevil (*Callosobruchus maculatus* and *chinensis*) we see an example of asymmetric sexual interference. Once again, males mate indiscriminately with females of both species. During copulation, males of the former do not have adverse effects on females of the latter, and the females can undertake subsequent copulations with ease. However, males of the latter do have negative effects on females of the former, rendering them unable to partake in further sexual activity. In a way, this scenario can be viewed as a kind of bridge between the highly adverse effects of some heterospecific copulations (as described above), and the ones that result in healthy (albeit sterile) offspring: the products of hybridization.

If the scene for heterospecific copulation was that of a horror film, the scene for hybridization is more like that of science fiction. In nature when we see successful hybridization, this means that species are so closely related that copulation, fertilization and gestation have all been successfully achieved. While this is certainly eerie and a subject of curiosity and intrigue, chances are that this high level of sexual success doesn't involve genital mutilation or a significant degree of bodily harm. Hybridization in nature occurs more frequently than most people think – across diverse animal phyla. For speciation to occur, organisms of the separating species must first be the same, and then somehow different. This means that historically there will have been an interval during which the two species were much closer than they are now. Successful interbreeding can occur, which leads to either viable or inviable hybrids depending on the extent of divergence that has taken place between them. This kind of grey area with respect to species boundaries can result in some fascinating behavioural patterns of the organisms exhibiting them.

For example, there are intense interspecific dominance relationships between two closely related chickadees in British Columbia, Canada. Black-capped (*Poecile atricapillus*) and mountain (*Poecile gambeli*) chickadees tend to be segregated due to differences in their ecology. Mountain chickadees prefer to live in high-elevation dry conifer forests, whereas black-caps prefer low-elevation mixed forests. However, thanks to the mixed forest-treatment practices of the logging industry, there are several areas where the two species overlap and create hybrids. Interestingly, there is a distinct directionality to the way that hybrids are formed. Male black-caps are much more dominant than their mountain-male counterparts, and this is a trait much desired by females. Hybrids are almost always formed from inter-species copulations between female mountain chickadees and male black-caps, meaning that females of both species prefer the more aggressive males. Luckily, the mountain men (not preferred) can still rely on environmental segregation in most of their range to make up for their lack of macho-man behaviour as compared to the black-caps.

In many cases where it occurs, the extent of hybridization

depends on the rate of encounter between members of the same species. When conspecific males are scarce, females are less choosy with respect to their mates. This makes sense, because having 'almost-the-right-sperm' is better than having no sperm at all. There are organisms, such as closely related species of swordtail fish, where interspecific hybrids are viable, although their fecundity is lower than that of pure breeds. Indeed, for most species that can experience hybridization there are behavioural and physiological barriers to prevent it. For the most part females will choose males of their own species; however, varying environmental and ecological factors mean that this isn't always possible.

Queer as fauna

It never ceases to astonish me when I have conversations with people who think that homosexuality is something unnatural or immoral. This kind of utter ignorance can result from societal and/or religious propaganda, and it's something that I wish people were better informed about. The honest to goodness truth of the matter is that you would be hard pressed to find a species on this planet that does NOT engage in some kind of homosexual activity. The utter ubiquity of homosexual sex that occurs on planet Earth means that if we are going to consider 'natural' behaviours as those that occur among animals in 'nature', then homosexuality is perfectly natural.

Upon initial consideration, the notion of homosexuality appears to be a paradoxical situation in terms of maximizing one's biological fitness. After all, sexual activities of any kind that don't directly attempt to create offspring are assumed to be biologically irrelevant. However, there are many plausible hypotheses for the existence of homosexuality that could account for an overall increase in the lifetime biological fitness of the organisms undertaking it. Before I delve into the many possible reasons for the prevalence of homosexuality in nature, it will be useful to explain a few basic terms related to it.

Same-sex behaviour is the most common form of homosexuality observed. It describes any kind of sexual behaviour towards members of the same sex, from courtship through to full copulation. *Same-sex preference* describes a situation where homosexual sex occurs despite the fact that heterosexual partners are available. Same-sex preference is much less common than same-sex behaviour, but it still occurs to a large extent. Organisms displaying both same-sex behaviour and same-sex preference also engage in heterosexual sex, and examples of these behaviours will form the bulk of this section. Lastly, *same-sex orientation* pertains to an exclusive preference for sexual partners of the same sex. This kind of behaviour is prevalent in humans, but also a few other mammals, like male chinstrap penguins (*Pygoscelis antarctica*) and male bighorn sheep (*Ovis canadensis*).

So how do we explain the apparent paradox of homosexual behaviour in the animal kingdom? As I mentioned above, there is no shortage of possible, biologically relevant reasons why a certain amount of homosexual behaviour may be adaptive and lead to an overall increase in an individual's biological fitness. That said, there are also a few explanations for homosexuality that suggest it to be maladaptive (i.e. with negative impacts on an individual's biological fitness). I'll begin with a discussion of the former before focusing on the latter. First, the *social glue hypothesis* suggests that homosexual behaviour may facilitate bonds and alliances between individuals, which could reduce overall tension and prevent future conflict. For example, in bottlenose dolphins (*Tursiops* species), alliance formation between males is an extremely important part of the mating process. They form either pairs or trios, and they work together to establish exclusive control of a particular female for a certain period of time (usually from one to six weeks). They protect her from outside males, and take turns copulating with her. These male:male coalitions are termed *consortships* and are often maintained over many years. Males from a few different consortships can pair together as an *alliance* in order to help each other maintain control over their respective females. Occasionally super alliances of up to fourteen individuals form as a way to have a 'team' of guys on your side to help defend your reproductive rights to a certain female. This kind of mating strategy means that it is extremely important for males to form strong social bonds, and engaging in sexual activity with each other is a logical way in which to accomplish this. Indeed, the incidence of homosexual activity between males is much higher than that observed for females, although both male and female bottlenose dolphins engage in a high level of homosexual sex compared to other mammals. In fact, juvenile male dolphins engage in a higher level of homosexual than heterosexual sex. Sexual contacts between males occur approximately 2.38 times per hour (*every* hour of the day), earning the biological description of hypersexual. Events are long in duration, and male partners frequently change roles, again suggesting that this behaviour cements social bonding as opposed to establishing dominance.

Homosexual interactions between male parrots suggest another form of social glue. Males utilize these close encounters to assess the physical condition of their partners with respect to their own. In other words, homosexual sex functions to solidify social positions within a group, and allows males a chance to decide whom they may or may not follow. Males chosen as preferred leaders may have superior foraging or risk-assessment skills, and are more likely to be followed by other males following a thorough homosexual encounter.

The social glue hypothesis also suggests that homosexual sex may be a means of facilitating reconciliation between individuals subsequent to a conflict. This is certainly the case with one of our closest primate relatives, the bonobo (*Pan paniscus*). Post-conflict copulation is normal for both male:male and female:female partnerships, suggesting it as an established means of decreasing tension subsequent to aggressive encounters. Homosexual encounters between male bonobos include kissing, fellatio and genital massage, whereas encounters between females include kissing and genito-genital rubbing. When new females migrate to an established group, homosexual contact with resident females is commonplace and viewed as a means by which to reduce angst associated with immigration. This activity serves to express cohesion and affiliation between females, providing further solid evidence in support of the social glue hypothesis.

It's worth keeping in mind that the ubiquity of homosexual sex that occurs in nature means that its existence is likely to be explained in different ways in different scenarios. Although the social glue hypothesis is well supported by the data from both bottlenose dolphins and bonobos, it doesn't always explain the situation as well as other hypotheses. For example, the *intrasexual conflict hypothesis* suggests that homosexual copulations occur as a byproduct of male:male competitiveness. Male vervet monkeys (*Chlorocebus pygerythrus*) exhibit behaviours that support this idea because they engage in homosexual activities in order to establish and reinforce dominance hierarchies, and they do so in rather dramatic fashion. Submissive males will retract both their testicles and their penis into their bodies as an illustration of their low status, while dominant males will keep both extended. Dominants will orient themselves in a perpendicular direction towards the submissives, as if to display their erect penises in a show of high-ranking prowess. They will also circle around submissive males, displaying their ano-genital regions. There have also been observed instances of inferior males submitting themselves to dominant ones, as if to reiterate their low-ranking status. Similar dominance sexual routines occur in many primate species. Silvery marmoset males engage in complicated homosexual dominant versus submissive routines that often see the dominant partner displaying an erection (as if to say: I am allowed to have this, you are not). Low-ranking males will crawl towards dominant ones, making meek vocalizations and smelling the dominant's genitals. A similar routine takes place in squirrel monkeys (*Saimiri* species), although the dominant male will often spurt some urine during the process. These examples lend support to the idea that there are matters of intrasexual (same-sex) conflict that may be remedied through a good old roll in the hay, although establishment of dominance hierarchies is not the only conflict between males that could result in homosexual copulation.

It's thought that males of many insect species engage in homosexual activities as a way to decrease the biological fitness of their male lovers. OK, so that sounds confusing on the surface, but it makes sense in the context of male:male competition. Whenever males engage in sex, there are two roles: the giver (the one doing the mounting and penetrating), and the receiver (the one being mounted and penetrated). Receiving males can modify their demeanour in order to manipulate other males into mating with them, most often by appearing as females. This kind of pseudo-female behaviour has been observed in several insect groups including termites (order Isoptera) and fruit flies (*Drosophila* species), and it can take the form of female sex pheromones that the males secrete to attract other males. In a few species of anole lizards (genus *Anolis*), subdominant (small) males pose as females so they can sneak into a dominant male's territory and get close to the females living there. These 'transvestite' males also receive copulations from the dominant males. The logical question that comes to mind is '*Why* would a male want to trick another male into having sex with him?' The answer makes a lot of sense from a biological perspective. In addition to reducing aggressive tendencies of dominant males, this kind of homosexual submission distracts aggressive givers from mating with females (which may leave more female mating options open for receiving males if giving males get 'spent'). Essentially, they may be tricking a dominant male into wasting his sperm on them, instead of on a receptive female. In other cases, receiving males could take advantage of the nuptial gift being provided by giving males, even stealing it for their own nutritional benefit.

Another potential reason for homosexual behaviour is to gain practice. Males of many species will not initially have either the social rank or the sexual skills to successfully woo a female, and in many cases a little homosexual practice goes a long way towards achieving these things. Male bovids (cattle) engage in such a large amount of homosexual mounting that it is considered to be a nuisance in the cattle industry. Homosexual activity in male bison (*Bison* species) supports the sexual practice hypothesis, as most interactions occur outside the breeding season. In addition, young individuals are more frequently involved, and sexual activity is also associated with playful behaviour. It's like combining a little hide-and-seek with a literal hide-and-seek.

A similar scenario is observed in langur monkeys (subfamily Colobinae) and several species of macaques (genus *Macaca*) where it is hypothesized that homosexual mountings facilitate social and motor development. In macaques, sufficient practice of sexual mounting as an immature individual is crucial to sexual success as an adult; the sex (male versus female) of the mounting partner in these juvenile copulations is inconsequential. Immature langurs, as young as three months of age, frequently engage in 'practice' copulations with mature partners of either sex, although the combinations of female:female and male:male are most common.

As with the immature bison, play is frequently associated with these practice sex sessions, and partners often prefer to copulate with others of the same age and size.

Now, what about sex merely for the sake of the act itself? After all, humans engage in homosexual and heterosexual activities completely outside of the realm of biological fitness. Biologists often get carried away with our requirement to find an evolutionary purpose for all animal behaviours, and while this makes good scientific sense for the most part, there are times when it doesn't. Sex is awesome. It would therefore be rather assumptive for us to conclude that something similar never occurs in other animals. Indeed it does. Male mountain gorillas (*Gorilla beringei beringei*) engage in a number of homosexual behaviours, such as genital stimulation, pelvic thrusting and ejaculation. Interestingly, this kind of behaviour in gorillas is classified as simply sexual, not socio-sexual (as most of the other examples described) in not having any kind of social reason for its existence. It's sex for the sake of sex. Male silverbacks will engage in homosexual encounters with subordinate and immature males, and will initiate these encounters with copulatory vocalizations. Silverbacks often experience ejaculation through these interactions, and an individual will often have preferred male 'lovers' from within his group that have sex only with him. Silverback males may also fight between themselves for access to specific subdominant male lovers, sustaining massive injuries in the process.

On a less violent note, there is an awful lot of 'just for fun' homosexual sex between female Japanese macaques (*Macaca fuscata*). They repeatedly mount preferred partners in bi-directional manner, utilizing several varied positions and engaging in genital rubbing, groping, head-bobbing and intense gazing into each other's eyes. Behavioural data for Japanese macaques does not support the parameters for any of the socio-sexual hypotheses described above; instead, it suggests that female:female courtship and sex follows a pattern of a purely sexual behaviour and mutual attraction between partners. In other words – some female Japanese macaques are just gay.

Thus far we've had a rousing discussion (not exhaustive) of the ways in which homosexual behaviour may be adaptive. However, there are times when homosexuality may not be adaptive. Sometimes, homosexual sex can be a result of 'something gone wrong' in the genetic or behavioural department. Again, several hypotheses may explain the existence of maladaptive homosexuality. First off, the *mistaken identity* hypothesis describes a common problem in the insect world, namely that males and females are virtually identical (except for their genitalia). Mistaken male:male sex is very common in organisms like fruit flies (genus *Drosophila*) and bat bugs (family Cimicidae), leading to the evolution of structures or behaviours that allow males to showcase their maleness to other males prior to copulation. In fruit flies, this is accomplished through the use of pheromonal cues, and in bat bugs males have evolved a secondary paragenital structure that is different from the ones possessed by females. Mistaken identity may also result when a male has previously had sex with a female and retained some of her sexual pheromones, resulting in homosexual courtship by other males.

In other cases in the invertebrate world, homosexuality is a byproduct of a behavioural syndrome, which means that a group of behaviours is genetically inherited – some adaptive and some not. Generally, when a group of behaviours forms a syndrome like this, the fitness benefits gained from the adaptive ones are much outweighed by the fitness detriments of the non-adaptive ones. For example, some lines of seed beetles (subfamily Bruchinae) have a genetic disposition to have a higher mating frequency than others. Individuals with these high mating frequencies have a higher incidence of homosexual interactions as a natural by-product of the fact that they are simply having more sex. Other genetic evidence comes from fruit flies, where modifying certain genes can increase same-sex interactions between males.

What about homosexual behaviour as a function of sex ratio? The so-called *prison effect* explains a situation in which the propensity for homosexuality is increased when finding a heterosexual partner is difficult. Female Laysan albatross (*Phoebastria immutabilis*) are more likely to form same-sex bonds when male partners are scarce. This represents a possible alternative strategy for these females to increase their biological fitness, since successful rearing of a chick requires the efforts of two parents. Females pairing together to raise offspring fare better than females who do not reproduce at all, and in addition, the male who fertilizes them enjoys the genetic benefit of his extra-pair copulation(s). Another example comes from Gouldian finches (*Erythrura gouldiae*). When males are experimentally housed in all-male groups, they form homosexual partnerships with strong social bonds. Once these bonds have been formed, introducing a female into the mix does not destroy them. Despite the fact that these males *could* form heterosexual partnerships, they do not. They remain in a homosexual relationship with their chosen male partner.

The human animal is certainly not exempt from the prison effect, seeing as the very name of the phenomenon reflects a situation where male and female inmates are kept in separate quarters. Humans are sexual creatures, and in the absence of heterosexual partners we will actively pursue the next best thing. In rural China, where the notorious 'one child policy' has resulted in an overabundance of bachelors (to the tune of 30 million), incidence of homosexual activity has increased. This is not surprising considering that the chances are very low that most of these extra men will ever find a female partner. The sex ratio is simply too skewed, and there will be millions of males who die without leaving any kind of biological footprint.

Although the short section above has focused on the so-called 'maladaptive' hypotheses for homosexuality, I want to stress that the bulk of evidence for these come from examples in invertebrates where individuals have difficulty identifying one sex over another. It's incredibly important for me to end this section on homosexuality by stressing that, for the most part, barring short-term skews to sex ratios, homosexual sex in vertebrates, mammals and primates can be clearly explained for its *adaptive* and *biologically important* functions in animal societies. Hear me loud and clear: homosexuality is ubiquitous, natural, and important for several aspects of heterosexual function in almost all animal societies.

Sexual coercion

There are times when I have to chuckle at us biologists for creating a separate set of terms to describe behaviours in 'other' animals as opposed to in humans. For example, sexual coercion is basically another term for rape. Yes, male animals rape female animals, but somehow it seems more politically correct to call it sexual coercion. The terms refer to an act of courtship or copulation that is aggressive by the instigator and unwelcome for the recipient. Essentially there are three basic requirements for a behaviour to be termed sexually coercive: first, it must directly increase the mating success of the aggressive male. Second, it must suppress the opportunity of the female to make a choice about her sexual partner, OR it must force the female to mate with a non-preferred male partner. Third, it must impose some kind of fitness cost on the female victim.

Unfortunately, sexual coercion in the animal kingdom is extremely common, and in most cases it has everything to do with the overlying theme of the value of eggs versus sperm. We've seen this theme emerge repeatedly, from ejaculate composition, deception and cryptic female choice. Here we will examine some of the strategies employed by mammalian and primate females to deal with the various modes of sexual coercion that they face.

Female Northern elephant seals (*Mirounga angustirostris*) breed synchronously in large harems that are guarded by a massive alpha male. A single aggressive male can have mating exclusivity over hundreds of females, who all gather in a specific breeding ground year after year. Once females have hauled themselves out of the water, they are frequently subjected to harassment from subdominant males, which is why they tend to gather in such large groups with a 'protective' alpha. Being in a large group with other females dilutes the level of harassment suffered by any one individual. In addition, females make vocal protests to almost any encroaching male, which often grabs the attention of the alpha. Termed the 'trade sex for protection' hypothesis, the alpha will interrupt or prevent sexual attempts by other males on ladies of his harem. This strategy is quite successful because mating disruptions are extremely common following protest vocalizations of victim females. Risk of harassment by subdominant males on breeding grounds can be so great that some females mate out at sea instead in the breeding colony. In a species like the elephant seals where the level of sexual dimorphism (i.e. males are significantly larger than females) is significant, it makes sense for females to do whatever they can to avoid violence and possible injury from overzealous suitors.

Orangutans (*Pongo* species) also exhibit an extreme amount of sexual dimorphism. Males are much larger than females, and can easily overpower them physically. However, the breeding structure in social groups is much different in orangutans than in elephant seals. There is still an alpha male, who is the largest and most dominant individual in the group. He has massive canines and large cheek flanges that signal his leading role to all others. Females mate with the alpha, but they also mate with other sexually mature males in the group. In fact, females generally copulate more than 500 times per conception. It's thought that they do so in order to appease the sexually coercive behaviour of the aggressive males, and also to create paternity uncertainty (as discussed in an earlier section).

There are several other primate species where sexual coercion is commonplace, most often where there is a large difference in size between males and females. In one of our closest relatives – the chimpanzees (*Pan troglodytes*), males tend to be most violent and coercive to females that are currently ovulating, which is an unfortunate consequence of the fact that females exhibit external sexual swellings during this phase. Female chimps cannot use a strategy like paternity confusion (as orangutans do) because there is no question about which sexual encounters are likely to create offspring. However, these females will still mate with multiple males during their most fertile phase, although the more sexually coercive males tend to achieve the highest success rate when it comes to the number of copulations. Females exhibit increased levels of glucocorticoid steroids concurrent with high levels of sexual coercion, indicating that this is a stress-inducing and costly event for them. Despite the short-term stress and injury caused by sexual coercion in female chimpanzees, it is hypothesized to result in an increased lifetime biological fitness because the most dominant males (i.e. the ones most capable of sexual coercion) contribute the greatest genetic benefits.

There are two kinds of aggression that female chimpanzees receive (primarily during ovulation, but also during the rest of the cycle). First, males use their physical power and coercion to control female resistance and force copulations. Females appease this behaviour by essentially 'offering themselves up' during ovulation to the males that have been most aggressive towards them during the rest of their cycle. Second, high-ranking males use coercion to constrain female promiscuity, which increases their own chances of fatherhood but provides almost no chance for females to confuse paternity. The scenario is almost completely different for our other closest primate relatives, the bonobos (*Pan paniscus*). Although there is some evidence of dominance relationships in both male and female bonobos, there is a complete absence of male aggression towards any females displaying signs of increased fertility.

Hamadryas baboons (*Papio hamadryas*) take the act of sexual coercion to another level. These primates live in small groups where males monopolize a semi-permanent harem of females. They guard 'their' females aggressively via herding, and females are generally compliant and sexually submissive to their leader. This is called female defence polygyny, indicating that the male defends his harem (and also keeps tight control over their sexual behaviours). Males will also fight with other males to defend their harem; however, takeovers by more aggressive males are a fairly common occurrence, especially for young males looking to begin their reproductive careers with a 'set' of fertile females. When a takeover happens, the losing male is forced to flee, and his resident females are forcibly and aggressively transferred to their new 'owner'. Paternity confusion is impossible for females in this social scenario, and when a takeover happens the new dominant male kills all the existing offspring. The act of infanticide is clearly advantageous for the new dominant male, who should avoid providing care to unrelated offspring. In addition, the abrupt halt to lactation brings his new ladies into oestrus, meaning that he can get to work creating some offspring of his own. Males that already have harems can also take over those of other males, thereby increasing the number of ladies at his service. They are especially aggressive towards newly acquired females, coercing them into compliance. One would imagine (and one would be

right) that this is an intensely stressful time for the newly transferred ladies. In addition to having all their offspring killed by
their new dominant male, they are the target of his extreme
aggression as well as contest competition for resources from the
other females already in his harem. Indeed, the inter-birth interval, the time between successful births per individual, is higher
for females that have changed hands than for females that have
not. In other words, there is a physiological cost to females that
comes along with being forcibly transferred between dominant
males and having their offspring killed. (No kidding!) Inter-birth
intervals of females remaining in the same harem after a takeover
are not increased, suggesting that takeovers are only stressful for
the newly acquired females.

Although the immediate costs of sexual coercion in any species
seem high for the female and low for the male, as I mentioned
earlier there could be benefits to the female's biological fitness in
the long term. The Hamadryas baboon scenario is indeed strange,
what with males having such extreme dominance over females;
but over the long term one can think of a female as being transferred up the dominance chain of males in the local population,
such that she eventually ends up with the most dominant aggressive male. He will have successfully taken over the harems of all
males before him, thereby proving his strength and power. On a
level of pure biological significance, it makes sense for a female to
mate as much as possible with this, the most aggressive male, so
that she can have offspring with his genetic characteristics. Taking
it to an anthropomorphic extreme, one conjures up an image of
an old woman being proud of her son – the most intense rapist
and murderer of the population.

Cannibalistic females and the males that love them

We've all heard the horror stories of sexual cannibalism in the animal world – they are titillatingly gruesome. Sexual cannibalism represents perhaps the most extreme form of conflict between the sexes, and it is far from being a simple endeavour. This kind of behaviour is observed in several species of praying mantids and spiders, and there is tremendous diversity in the extent to which it occurs. Let's begin with a little about the basics of eating one's mate. Females have much to gain (and potentially much to lose) by being sexually cannibalistic. If she plays it right a female can enjoy the benefits of a male's genetic donation, his nutritional nuptial gift, and then, his entire body. However, if she takes her cannibalistic tendencies too far she may risk not finding a future mate, so there is a set of factors to be considered any time there is the potential for cannibalism. Circumstances including the state of current resource availability (either food resources or male resources) and the quality of said resources are critically important.

Ideally, a female should be able to weigh all factors involved and make appropriate decisions based on ever-changing ambient conditions.

In reality, this seldom happens. In fact, it's quite common for females to cannibalize males before *any* copulation has even taken place. So-called pre-copulatory sexual cannibalism is rampant and there is a substantial set of hypotheses to explain its existence. First, during times of severe food shortage, it makes sense for females to partake in cannibalism to ensure they have enough to eat. In most sexually cannibalistic species females are substantially larger than males, so catching, subduing and ingesting them is not a problem. Second, a female may want to remove undesirable males from the gene pool in order to prevent herself from mating with them and potentially lowering her own biological fitness. Third, the possibility exists that females may misidentify males of their own species, treating them as dinner rather than as sex partners. Several species of predatory spiders prey on other spiders with similar physical, behavioural, environmental and chemical aspects of their biology, and so it's not entirely surprising that a female may get it 'wrong' from time to time – chomping down on a conspecific male instead of mating with him. A fourth and much studied reason for pre-copulatory sexual cannibalism is the existence of *aggressive spillover*. The 'aggressive spillover hypothesis' (ASH) posits that some females are genetically predisposed to indiscriminate aggression, which results in them being formidable predators (and consequently makes them quite large and fecund). However, the aggression that makes them such effective predators also makes them particularly aggressive towards potential male suitors – and results in murderous rampages instead of courtship and copulation. Since the aggression is part of a 'suite' of behaviours, this is called a behavioural syndrome, and it means that this genetic disposition can carry with it both positive and negative effects.

In the case of the funnel web spider *Agelenopsis pennsylanica*, there is evidence for the nutritionally based hypotheses for pre-copulatory sexual cannibalism. Females that engage in it are more likely to produce egg cases than those who don't, and said egg cases are more likely to successfully hatch than those of non-cannibals. In fact it's been documented that sexual cannibalism increases a female's attractiveness to a male. It seems counterintuitive for a male to prefer to mate with a cannibal – but on closer examination of the facts it actually does make biological sense. Approximately 38 per cent of the time, females cannibalize their first 'mate' prior to copulation. However, this number drops substantially for future copulations – only 5 per cent of females undertake a second victim prior to mating with him. Biologically speaking, a male would be making a smart choice to pair up with a cannibal because chances are she's not going to cannibalize again, and since she has had a nutritious meal (i.e. the first mate), her chances of laying a viable egg case are substantially increased. So as long as a male isn't the first one to tap into an aggressive female, he's setting himself up well to try to mate with a cannibal. As far as the female is concerned, it appears as though the first male to darken her doorway is nothing more than a snack.

Burrowing wolf spider (family Lycosidae) females seem to conform to the tenets of the aggressive spillover hypothesis. Large, fecund females are much more aggressive than smaller ones. Aside from being voracious (successful) predators, they also engage in pre-copulatory sexual cannibalism. Large, aggressive females are not choosy about the males with whom they copulate. When given a choice between males of different phenotypes (both high and low quality), the dominant ladies show no preference. Whereas smaller females (i.e. those without the genetic predisposition towards aggressiveness that the ASH proposes) do exert selective pressure on the male population by killing and eating males in poor condition, and mating with those in superior condition. In this way, the weaker females are the ones shaping the phenotypic evolution of the male population, while the aggressive ones are simply killing and/or mating at will.

On the surface it may seem as though females are in complete control and the poor males who show up for sex are subjected to their desires, but this is not always the case. In addition to there being instances where males do exert some control over their survival status, there are also cases of reversed sexual cannibalism. The beauty of sexual selection and coevolution means that males of some species have developed counter strategies to combat their impending doom. That said, it's not necessarily a bad thing to be cannibalized. Although it's a stretch for the human male to envision when it would ever be strategically advantageous to allow himself to be killed and consumed, for several species of spiders and mantids who will only ever get one or two chances at reproducing anyway, putting all of one's reproductive potential into one or two matings makes a lot of sense.

Some male spiders, like Australian red backs (*Latrodectus hasselti*) perform a specialized sexual 'somersault' into the awaiting jaws of their female lovers so as to facilitate their own cannibalistic death. They position their abdomens directly above the female's jaws, and the ladies typically nibble on their posterior bits while copulation occurs. Optimally, a male red back would like to mate twice. He has two pedipalps (sexual organs) which can only be used once each, and females have paired sperm storage organs which require two copulations to fill. To die after one bout of copulation (as happens in approximately 12.5 per cent of the population) would almost be a cop-out for a male, since not only is he physically built with two pedipalps, he also significantly increases his biological fitness if he can utilize them both. Males have therefore developed a nifty technique to minimize their chances of succumbing to a cannibalistic demise during the initial bout of copulation. They constrict their abdomens, thus pushing their vital organs to the anterior of the body where they are protected from the female's nuptial nibbles. Males that undertake this constricting strategy minimize the damage to their bodies, yet still experience a successful copulation. They are then able to fulfil their biological destiny by engaging in a second bout of copulation. Finally, with both pedipalps spent, they allow themselves to be killed and consumed by their second female partner.

We see similar strategies in other spider species, like the orb-weaving *Nephilengys malabarensis*. In this species males engage in 'remote' copulation with females, effectively transforming themselves into functional eunuchs. Remote copulation? More like remote control. You can have sex with your partner while sitting comfortably on the couch across the room. Males of many species will break off a part of their sperm transfer organs (pedipalps) inside the genital opening of a female to serve as a mating plug and prevent her (to varying degrees of success) from receiving sperm from other potential suitors. Male orb weavers take this a step further. By breaking off the entire appendage, sperm transfer to the female continues once the male has made his exit. Not only do fully severed palps (penises) increase the total amount of sperm transferred to a female, they also increase the rate at which transfer takes place. In this way, an emasculated male can escape with his life and enjoy substantial reproductive success. Of course, it only makes biological sense to do so while he still has one functioning pedipalp – because once both are gone his chances of increasing his biological fitness are effectively zero.

So what else can a male do to avoid (at least initially) the awaiting clutches of a cannibalistic female? In the case of the nuptial-gift-giving spider *Pisaura mirabilis,* he can feign his own death before she has a chance to kill him. Death feigning (or thanatosis) is a common strategy for predator avoidance, but it is also observed in extreme cases of sexual cannibalism. In these spiders, males provide females with a nuptial gift. While she is busy unwrapping and consuming the gift, the male is busy copulating and transferring sperm. Females will often interrupt copulation several times prior to full sperm transfer, and it's during these pauses that the males play dead. He will collapse and remain completely motionless while retaining his hold on the gift. Once she's gone back to enjoying her snack, they will carefully 'awaken' and resume the process of sperm transfer. It's been shown that the males who play dead score a significantly higher number of copulations with females, as well as experiencing longer bouts of sperm transfer.

Not all aspects of male behaviour in response to female sexual cannibalism are as extreme. There are many ways in which males can maximize their biological fitness while copulating with cannibalistic females, especially in species where breeding is not limited to a short season or a few copulations because there is a direct conflict between males (who would rather live and mate again) and females (who would rather cannibalize them). Males are inherently aware of their surroundings, and can utilize the power of observation in order to avoid being killed. For example, male Chinese mantids (*Tenodera sinensis*) and Australian mantids (*Pseudomantis albofimbriata*) will cautiously approach females and attempt copulation from a position on her back, preferably while she is busy grooming or feeding. Male black widow spiders (*Latrodectus* species) will court well-fed females over ones that have been starved by using chemosensory cues from the silk on her web. By investigating the feeding history of a potential mate, the males can save their own lives by copulating with her only when she's full. Clever!

Males can also vary their reproductive strategies based on their ambient environment. In some cases, it may make more sense to self-sacrifice; in others, not so much. There is no shortage of examples to describe the intricacies of when it may or may not make sense for a male to self-sacrifice. Male orb weaving spiders (*Argiope bruennichi*) are more inclined to sacrifice themselves if their sexual partner is a virgin. This is likely beneficial considering that, if she hasn't had any other mating partners, his sperm will probably be the winning seed. In order to protect his assets, he should ideally monopolize her attention for as long as possible, and the best way to do this is to keep her busy devouring his own body. Male orb weavers are less likely to be self-sacrificial when their mating partners are not virgins, making a greater effort to escape her cannibalistic clutches if they don't have a monopoly over her eggs. In addition, males are also more likely to sacrifice themselves to unrelated females (as opposed to siblings). These orb weavers are fairly sedentary, meaning that the incidence of inbreeding can be quite high. However, the extent of genetic relatedness plays a role in the level of 'desirability' of any given female (high relatedness being less desirable), and males tend to prolong the duration of copulation and be cannibalized more frequently when paired with a non-sibling mate.

Male mantids (*Tenodera sinensis*) also have the ability to control self-sacrifice depending on their encounter rate with females. If ladies are hard to find, males are more apt to take higher risks to mate with them. Mantid males are also more likely to approach potential partners in windy conditions, as females have a more difficult time detecting them on wind-blown leaves. Males are prone to throw caution to the wind (pun intended) and approach a female more quickly and directly when their substrate is moving and they are more difficult to spot.

Although the majority of spider species have females that are markedly larger and more cannibalistic towards males, there are a few cases where the opposite is true. For example, sand-dwelling wolf spiders (*Allocosa brasiliensis*) are sex-role reversed, meaning that males have a larger investment in sexual reproduction. Unlike many spiders, these males do a lot more than simply contribute their DNA. They construct large, deep burrows and remain sedentary within them while the females are mobile and initiate courtship. Once copulation has taken place, the male exits the burrow and seals the female inside. She will emerge approximately one month later to disperse the spiderlings. There are a number of interesting aspects to this mating system. First, males are larger than females, which is quite the opposite of the general trend of size dimorphism observed in spiders. Second, the construction of the burrow represents a large parental investment by the male. Third, the female is able to mate more than once during her lifetime, although the hatching success of the first egg sac is highest. These factors set the scene for a rare occurrence in the animal kingdom: reversed sexual cannibalism. The male is more likely to cannibalize a female that has already mated, which makes sense for both his nutritional well-being and the fact that already-mated females have a lower reproductive success. In the few species in which male sexual cannibalism is observed, the trend is to eat the older, more vulnerable females, and to mate with virgins.

I'd like to close off the discussion of sexual cannibalism with a remarkable example of the beauty of evolution. In this case the result is far from beautiful, but it brilliantly represents the power and efficacy of sexual selection. There is only one native praying mantid species in New Zealand (*Orthodera novaezealandiae*). The unfortunate introduction of the South African praying mantid (*Miomantis caffra*) has resulted in displacement of the native species. As the invasive species increases its range, populations of the native species are declining to the point of local extinction. Dynamics between introduced and native species are often such that introduced organisms, which are usually able to withstand a variety of environmental conditions (making them successful invaders in the first place) and also in a place without natural predators, are able to more effectively compete for resources and wreak havoc on natural populations. However, in the case of the praying mantids in New Zealand, there is a more sinister reason for the demise of the native species. The sexual pheromones emitted by females of the South African species mimic the bouquets of the natives. The native males are more attracted to the introduced females than those of their own species. As if it isn't bad enough that the males are wasting their time and energy in courting the wrong females the scenario gets worse. Introduced females lure the lovestruck, unsuspecting males into their clutches and cannibalize them before they've had a chance to mate at all. The native males are lured away from native females to their untimely deaths at the hands of the invasives. Not only does this obliterate the reproductive success of the native species, the introduced females are obtaining a great deal of extra nutrition from the native males that they can then utilize to propagate their own species. Talk about a double insult.

Chastity belts

The human version of a chastity belt is a remarkable contraption made of metal and leather and, most importantly, a lock. The idea behind such a horrific item is that when eighteenth-century soldiers went off to war, they required a way to keep their ladies chaste. These hideous gadgets were fastened on to women's nether regions to prevent intromission of foreign penises while their partners were away.

Blockading the genitals of one's partner is of course a pretty unnatural practice for the human animal, but humans are not generally known for following the principles laid out by the notion of biological fitness. The use of chastity belts in the animal kingdom is widespread and extremely common for a number of biologically relevant reasons. The most common one is that after a male has mated with a female, it's in his best interest to prevent other males from doing the same, since this will mean that his sperm has the greatest chance of fertilizing her eggs. How best can he accomplish this? By inserting a mating plug either within or on top of the female's genital opening. A chastity belt for which there is no lock or key, just a biological barrier. Mating plugs are found in almost all groups of animals, from tiny nematodes to some of our closest primate relatives. Due to the often-contrasting needs of males versus females, the behavioural and physiological strategies with respect to mating plugs are as diverse as the organisms utilizing them. As if having your genitals gummed up with some kind of sticky substance wasn't enough, many mating plugs also contain appeasement substances that serve to decrease the sexual receptivity of females. Thus females are rendered physically and emotionally unavailable.

Mating plugs can also serve as a warning sign to other males. Having a large, morphologically obvious plug attached to her genitals sends a clear message about a female's mating status to would-be suitors, and depending on the particular ecology of the species at hand, it may result in males steering clear of 'marked' females. Based on these characteristics, the whole notion of mate plugging seems horribly unfair to females. There are some aspects of female empowerment when it comes to the formation of mating plugs, which we will get to later. Unfortunately, for the most part, the biology of the chastity belt is largely male-dominated.

The majority of mating plugs are comprised of amorphous material secreted along with other substances in the male's ejaculate. The blobby gel-like plugs start off in liquid form, and either infiltrate the female's vagina directly, or form a cap over the entrance that will harden over a period of hours or days. Some plugs are more effective than others, and it's been demonstrated with dwarf spiders that both age and size of the mating plug play a role in its robustness. Large plugs that have enough time to harden are much more effective than smaller ones or new ones that have not yet coagulated. Depending on the species, a mating plug of this sort can last from a few days up to an entire lifetime. Aspects of genital morphology have a lot to do with the lifespan of the plug. For example, most female spiders have three genital openings

– two for sperm deposit and a separate one for oviposition (laying eggs). This is an important characteristic of many invertebrate mating systems, because it means that a male can plug up a female without preventing her from laying his fertilized offspring. In most mammals and primates, sperm deposition and either oviposition or viviparous (live) birth must occur from the same opening – which means that mating plugs have a much shorter lifespan, dissolving in order for the next step of the process to successfully take place. In a perfect world (for the depositor of the plug), these chastity belts would stay in place right up until the last moment, however this rarely happens. Both males and females (depending on the species) have been observed to dislodge mating plugs left by others. A female may not be entirely happy with her copulatory partner, or she may have been forced into sexual activity. The last thing she wants is to be sealed up and stuck with offspring sired by an undesirable bachelor. Female ground squirrels (in the family Sciuridae) are known to 'groom out' mating plugs and consume them. In these and other rodents, primates, marsupials and bats, mating plugs are proteinaceous – so why not engage in a little post-coital snack?

It's not just females that would have an interest in removing mating plugs: males who have not yet had a chance to make a sperm deposit into a particular female will also have interest in removing her barrier. In some species, males spend a great deal of time and effort extracting mating plugs of others – only to then deposit one of their own.

Rats (superfamily Muroidea) have a well-studied mating system that often involves males becoming sexually satiated. This means that they literally get tapped out of sperm and it takes approximately fifteen days for their fertility to be completely restored. Various ejaculate parameters recover slowly, including components of seminal plugs and the semen itself. Interestingly, male rats exhibit something called the Coolidge effect, whereby exposure to a novel female stimulates sexual behaviour, even in sexually satiated males. So despite the fact he's got no semen to deposit into novel females, the male will go ahead and copulate with them anyway because it's still possible for him to mount and insert his penis into a female's vagina. He just doesn't ejaculate. Now, why would males engage in such behaviour when it's energetically wasteful and he has no chance of siring any offspring? You guessed it – in order to remove mating plugs that were deposited by previous copulatory partners. It's been found experimentally that these 'non-seminal' copulations actually function to dislodge the mating plug and thereby remove competitors' sperm. A mechanism like this may be why we see something called 'post-ejaculatory mounting' in male ring-tailed lemurs (*Lemur catta*). Once a male has finished copulating with a female, he will often climb back on to a mounting stance and grasp the female's torso without any kind of genital contact. It's thought that they do this to extend the lifetime of the copulatory plug by ensuring that it fully hardens subsequent to being secreted. While this may seem like a bad deal for the female, there are ecological circumstances where it's beneficial for her to engage in post-ejaculatory mounting with a male. For example, if he is a preferred partner, she may welcome the chance to further his reproductive success in this way. Indeed, females are often observed to solicit post-ejaculatory mounting from favoured partners.

The news does get better for females in general. Although males are responsible for producing the vast majority of mating plugs, there are several well-documented cases where females are the ones plugging themselves up subsequent to copulation. Although it may initially seem counterintuitive, there are many biologically relevant reasons for her to do so. In mating systems where males violently or coercively copulate, females can prevent unwanted sexual advances by being sealed off. In species where sex may not be violent but where a female has simply received enough sperm, it also makes sense for her to be blocked off so that she can spend her time and energy on other important activities. Giant wood spider (*Nephila clavipes*) females are solely responsible for production of a hardened plate over their genital openings. They do so in concert with oviposition, and these amorphous plates are highly effective at preventing subsequent copulations from (much smaller) males. Males are virtually unable to insert their pedipalps once she's put on her chastity belt. The only females who seal themselves up are the ones that have already repeatedly copulated, suggesting that once they have collected enough sperm it makes more biological sense for them to pre-emptively shut down advances from further suitors. Other females put their copulatory plugs to more sinister use. Females of an orb weaving species (*Leucauge argyra*) secrete a sticky substance from their genital ducts subsequent to copulation that functions to entrap males by their pedipalps. Males have to put up a struggle to free themselves (picture a spider-sized sticky trap), which gives the female ample time and opportunity to cannibalize them.

There are also cases where both males and females contribute to the production of a mating plug, which means that females exercise some kind of control over their choice of successful sires. Another species of female orb weaving spider (*Leucauge mariana* – closely related to the sadistic sticky trap ladies above) exercises one of two behavioural tactics during copulation. In some cases the female adds a clear liquid to the small blobs of white (mating plug precursor) paste deposited on to her genital plate by the male. These two substances react with one another upon mixing to form a smooth, solid mating plug that is effective at preventing further copulations. In other cases, the female does not add this 'magic ingredient' to the male's mating plug precursor, and no mating plug is formed. In the latter case the plug precursor material deposited by the male is simply dislodged or wiped away. So what is it that sets some males apart from others? It seems to be all about how good a job he's doing at stimulating the female during copulation. Something called 'copulatory courtship' refers to the specific movements of the genitals during copulation. The female orb weaver prefers a male that engages in rhythmic pushing on her legs with his, and that makes repeated short insertions of his genitalia into hers. When males perform these specific copulatory behaviours, females are much more likely to invest in creating a copulatory plug.

So far the discussion has focused on examples of mating plugs that are essentially glandular secretions of amorphous, proteinaceous stuff. Indeed, the majority of mating plugs fit this profile, regardless of whether they are produced by males or females. However, there are other kinds. In another variety, the sperm itself serves as a mating plug. Male scorpions (order Scorpiones) produce two kinds of mating plugs, but the more complex ones are comprised directly of sperm in various forms. During copulation the male inserts a gel-plug into a female's genital opening that is initially fluid. Once inside the female's reproductive tract it will harden in approximately three to five days. The main component of the entire plug is sperm. The outermost part of the plug (which is exposed to the outside world) is composed of inactive, tightly wound sperm packets. In the middle area the sperm begin the process of uncoiling from their tightly woven state, so that by the time they reach the innermost part of the plug (inside the female) they are free-swimming individuals. Some aspect of the fluid conditions within the female's reproductive tract activates the sperm from within the plug, although the exact chemistry is as yet unknown. These mating plugs are highly effective, remaining in place for up to two weeks. Since they are almost entirely composed of sperm, it could technically be described as a spermatophore (sperm packet), but the key difference between the mating plug of the scorpion and the deposition of spermatophores in so many other invertebrates is that it functions to block the female's genital opening.

Another organism that utilizes a sperm-filled mating plug is the red-sided garter snake (*Thamnophis sirtalis parietalis*). I've talked about aspects of their huge sex orgies in other sections of this book, but I'll just quickly recap. These organisms live in the extremely cold winter climates of central Canada, and they hibernate in underground burrows during the winter months. Upon waking from hibernation they engage in massive amounts of sex in extremely large groups (tens of thousands of individuals). Aside from being a major spectacle and tourist attraction, the unique mating behaviours of the red-sided garter allows us insight into aspects of mating plugs in reptiles. It's traditionally been assumed that the gelatinous plugs males insert into the genital openings of females serve to deter reproductive advances of subsequent suitors (the 'chastity enforcement' hypothesis). However, when males are surgically altered such that they are unable to create mating plugs, their sperm leaks out of the female's vagina almost immediately, indicating that the snake plugs also serve to keep sperm from escaping. Furthermore sperm are located throughout the plug, allowing for their gradual release into a female. The bottom line is that mating plugs can serve more than one function. Other important, non-exclusive functions can also be present.

The prevention of further copulations is certainly the most appropriate hypothesis to describe the function of the third and final kind of mating plug. In this case what's plugging up the female genitalia are fragments of broken-off male genitalia. It's quite a common occurrence in the Arthropod phylum, especially in spiders and other arachnids where males experience sexual cannibalism. They will literally break off their own genital structures within the openings of females to prevent intromission by other males. Usually this consists of parts or the whole pedipalp, the structure used by males to transfer sperm. I say 'usually' because there are a few examples where the male will utilize his entire (dead) body as a mating plug – although this is rare and its success at preventing future copulations is not yet clear. Males have two pedipalps, and so essentially have two chances at mating (and plugging) a female with them, as we have seen.

Perhaps the most titillatingly horrifying story comes from spiders of the *Tidarren* genus. Male *Tidarren* spiders have very large pedipalps that are highly effective at sperm transfer. Unfortunately, these large appendages are so heavy that males have a tough time getting around with their massive genitals in tow. Physical stamina of males is severely limited, which is bad news because there is a good deal of male:male interaction and mate-guarding in these species. Essentially, this amounts to a functional conflict. On one hand, large pedipalps enable males to be reproductively successful. On the other hand, large pedipalps make them poor fighters and poor mate-guarders. How are males supposed to manage such a massive conflict? The 'gloves off' hypothesis describes the commonly observed behaviour in *Tidarren* species where males emasculate themselves. They rip off their own members, transforming themselves into either half-eunuchs or full eunuchs. The removed members are then stabbed into females, and there they remain. Once they've rid themselves of their massive genitals, their body weight is much more manageable and they are much more effective at fighting and mate-guarding. Despite having no genitals left, males can still realize biological fitness by protecting their 'impregnated' ladies. These spiders are not sexually cannibalistic, so palp removal essentially amounts to *Tidarren* males having their cake and eating it too.

Breakage of male genitalia inside females may not *always* be under strict male control. For example, in the St Andrews Cross spider (*Argiope keyserlingi*), females appear to regulate a male's genital mutilation. Males are much smaller, and females forcibly terminate copulation by attacking them and breaking off their pedipalps. Females prefer large males to small ones. When experimentally mated to small males they terminate copulation much sooner than they do with large ones. As I mentioned, cessation of copulation is violent and results in breakage of the male genitalia; however, when it happens earlier, only small genital fragments are left behind because full penetration has not yet occurred. Therefore, small males are forced into leaving ineffective plugs, and females are free to carry on their copulatory activities with more desirable partners. They allow large males to copulate with them for much longer periods, which means that their pedipalps (although larger to begin with) reach deeper into the genital opening. In this case when she violently rips his pedipalp off, it forms a persistent mating plug that is able to withstand further copulatory attempts from other males. While it appears as though males are in control of the use of their own pedipalps for sperm transfer, this example demonstrates that, in nature, there is *always* an exception to the rule.

Fifty shades of BDSM in the animal kingdom

Sex is not always easy. In fact, carefree sex under the umbrella of romance and camaraderie is something that our species is virtually alone in enjoying. Although there is much give-and-take when it comes to the opposing needs of males and females in the animal kingdom, there are few scenarios more drastic than traumatic mating. Yes, it's actually termed traumatic in the scientific literature, and for good reason. Traumatic mating has evolved repeatedly in several animal phyla, although it is most common among arthropods (invertebrates with exoskeletons). Many kinds of behaviours fall into the general category of trauma, starting with those that involve non-genital structures that serve purposes other than those of direct sperm transfer. The technical term here (which provides a delightful description) is 'extra genital wounding structures', and it aptly describes the fact that these appendages serve to secure sexual partnerships with unwilling participants.

Take the love darts possessed by several species of terrestrial slugs and snails (phylum Gastropoda) (discussed in detail in 'When a Boy is a Girl Too'). These organisms are hermaphroditic, but they don't necessarily play male and female roles simultaneously. Love darts are calcareous structures that are literally stabbed into the body cavity of a potential mating partner, and they carry with them hormones that stimulate the female sexual role of the partner who is 'stabbed'. Evolutionarily speaking, it makes sense for a hermaphrodite to maximize its encounters as a male, because the female role carries a much higher reproductive cost. That cost is higher still when victims are 'accidentally' stabbed in the head, or through vital organs that cause serious injury or death. Love darts can be large and menacing (up to 30 per cent of the length of the snail's foot in some species), so they certainly live up to their name as extra genital wounding structures.

Although not as menacingly named, copulatory setae in earthworms function in a similar way. These organisms are also hermaphroditic, and unlike many gastropods individuals exchange sperm reciprocally and simultaneously. There are between forty and forty-four small, sharp spines located on the ventral sides of individuals that pierce through the outer skin of mating partners during the 3.6-*hour* copulatory process. Tissue damage caused by injection of the spines can be substantial, but the benefits gained by the 'injecting individual' are significant. The setae inject a substance into the receiving individual that instigates a refractory period – or a period of reduced sexual activity. This allows for a maximum amount of sperm to be taken up by the receiver. Similarly, male scorpions (order Scorpiones, not hermaphrodites) will inject their mates with several 'sexual stings' containing a pre-venom that is thought to modify her muscular contractions, like a form of tranquilizer. Although male wolf spiders do not appear to secrete any kind of substance into their female partners during

coercive mating, their massive fangs serve to keep them stead-
fastly secured, leading to longer copulation durations. Females
suffer loss of haemolymph (body fluid) and significant scar-tissue
formation. Traumatic indeed. We have yet to reach the point of
insemination, and we are already dealing with anchoring, wounding, secretions and more.

Before we move on to actual sperm transfer, it's worth noting
that all kinds of substances are traumatically secreted into females. There are those that induce refractory periods in receiving
females such that they refrain from immediate sexual activities
with other partners. This has the ultimate effect of allowing
the sperm they have already received to have a greater chance
of reaching the ovaries. There are also chemicals that decrease
sperm digestion within the female reproductive tract, that increase female receptivity to the currently mating male, and that
increase short-term female fecundity by increasing egg number
or overall egg size. Such chemicals are all utilized in exhaustive
combinations. Substances may be injected directly with sperm
or as a separate traumatic secretion transfer. The bottom line is
that there is much more than meets the eye when it comes to
traumatic mating and fluid exchange.

Perhaps the most well-studied examples of traumatic insemination come from the world of bed bugs (*Cimex lectularius*).
Instead of using their sword-like penises (parameres) to deposit
sperm into a female's genital opening, male bedbugs of many
different species will simply stab her outer body cavity and deposit sperm directly into her haemolymph. Sperm will then
migrate to her ovaries. In many species, females no longer have a
primary genital opening – why keep it around if it's never used?
A similar scenario plays out during sexual relations in the spider
Harpetica sadistica (yes, that's actually its official scientific name).
Males have a needle-like penile organ and females have atrophied

spermathecae (traditional sperm storage organs). Female bed-bugs have evolved a secondary genital structure (the spermalege) to replace the genital openings of female bedbugs of previous eras. The spermalege is officially termed a 'paragenital' structure, it is uniquely modified to allow stab wounds to heal more quickly and effectively than other areas of the cuticle. This is good news seeing as multiple stab wounds can have any number of negative effects on females' health, including water loss. Bed bugs are well adapted to survive for extended periods in fairly dry conditions found in human homes; however, females that are multiply mated and left with gaping wounds lose water almost 30 per cent faster than those without. The percentage of water loss becomes even higher and more life threatening when males are careless with their insemination, missing the spermalege and stabbing her in less well-protected areas. A few species of female bedbug have evolved to have two spermaleges – one on either side of the body. This makes good sense in terms of both wound healing and over-all fertilization success.

Bedbug males aren't as bad as others when it comes to the lo-cation of their sexual stabbing. Some females receive wounds at random locations on their bodies, from their abdomens to their eyeballs. Male sea slugs (*Siphopteran* species) stab females with their sharp needle-like penises directly into their foreheads. Sea slugs are hermaphroditic – the individual who does the stabbing is more likely to take on the male role during a particular mating. Although piercing your partner's forehead during sex may at first seem absurd (and gives completely new meaning to the phrase 'Not tonight, honey, I have a headache'), it may in fact be a re-markable demonstration of the power of evolution. The location of stab wounds between individuals is remarkably consistent to a region directly above the central nervous system, which suggests that transferred secretions may have a sophisticated and complex reaction with neural ganglia (also discussed in 'When a Boy is a Girl Too'). This kind of neurophysiological manipulation could act in a male's favour by influencing female behaviour in such a way as to maximize his reproductive success – something that has yet to be verified.

Sperm or other male-secreted substances need not be stabbed into a female in order for them to be traumatic. Consider the reproductive strategy of the nudibranch (sea slug) *Aeolidiella glauca*: males deposit a spermatophore (sperm package) on to the female's dorsal side (her back). How does the sperm manage to find its way into her body? Through a delightful process called histolysis, essentially chemical burning, where compounds in the spermatophore break down the outer tissues of the female and allow for the successful passage of sperm. Females are left with dissolved and decayed tissues that require a substantial amount of time and energy to rebuild. Sperm transfer via histolysis is observed in several organisms, and as with any other wound inflicted by traumatic processes, the mix of open lesions and sexual products provides serious potential for infection.

Whereas in female bedbugs the spermalege acts to minimize the risk of infection by expediting wound closure and healing, other females have evolved different strategies to combat infection. As we have seen, sexually transmitted infections are common in insects: they are highly pathogenic and play important roles in the evolution and ecology of their hosts. This is evident in many species that experience traumatic insemination, due to the fact that wounding and sexual activity are combined. Traumatic mating is common in many fruit fly (*Drosophila*) species, and it has been documented that females of some species up-regulate genes associated with immune function in response to male courtship song, kick-starting their immune systems prior to any kind of engagement with courting males. This functions to shorten the time delay between copulation and appropriate physiological response.

With all these examples of violent, traumatic, stabby behaviour, an obvious question springs to mind: why would a male evolve to inflict such pain on his mates? Apart from the obvious evolutionary conflicts between males and females, traumatic mating provides males with some distinct advantages that they wouldn't otherwise get. First, the many kinds of chemical compounds that males inject into females may be more effective when delivered traumatically rather than traditionally because females may have already evolved mechanisms to diffuse said compounds, and therefore mitigate their effects, when they are delivered through the reproductive tract. In addition, the reproductive systems of many females contain compounds that digest or resorb sperm. By completely avoiding these areas and depositing sperm elsewhere, a male can avoid such a sad fate for his precious seeds. Second, wounding a female during the mating process may prove to be advantageous to a male if the female must then utilize the bulk of her energy on wound healing as opposed to further mating. Third, in rare cases males may trigger a 'terminal investment' response in females, whereby females actively increase their sexual investment with their current partner knowing that she's unlikely to survive future reproduction. Overall, it is clear that there are situations in which traumatic mating, drastic and nasty as it is, makes perfect sense. There are also cases where it does not seem to make sense, like when males engage in traumatic mating behaviour with each other. A closer look at the intricacies of homosexual traumatic insemination may provide some insight as to whether or not this is adaptive, or simply a way to make the best of a bad situation.

Homosexuality is nearly universal in the animal kingdom, so it should come as no surprise that in species where traumatic mating is the norm, it occurs between males too. In male bed-bugs homosexual interactions are prevalent. In fact, male sexual interest is directed towards any newly fed individual, regardless of sex. The bloated, blood-filled belly of a large (and presumably healthy and fecund) individual is what gets his attention. It is only after mounting and physical examination that sexual identity can be confirmed. Considerable physical damage can be inflicted via sexual stab wounds as we have already learned; however, male bedbugs, unlike the females, have not evolved the secondary par-agenitalia (spermalege) to help heal their wounds more quickly, making male stab wounds even more serious. So what's one male going to do to prevent being stabbed by another? Newly fed males utilize their alarm pheromones to signal their sexual identity to approaching sexual predators. When such compounds are released, the chances of being mounted by another male are significantly decreased. Essentially, alarm pheromones are an ef-fective way of saying 'HEY! I'm a dude!'

Male African bat bugs (*Afrocimex constrictus*) take the concept of homosexual traumatic insemination to an astonishing extreme. In this species being stabbed by a sex-crazed male is common for both genders – and unlike the bed bugs, there is no pheromonal cue that allows males to ward off the advances of approaching suitors. Instead, males have evolved female-like paragenitalia. Yes, you read that correctly: males have evolved to have a structure that helps to protect their tissues from the unwavering traumatic stab-bings of other males. Interestingly, the paragenitalia of male bat bugs is structurally different from that of females, which makes sense considering that if a male evolved paragenitalia that was identical to females this would likely result in a greater number of unwanted homosexual assaults. Even more bizarrely, a significant chunk of the female population has evolved to have paragenitalia that resembles the more masculine form (i.e. females have evolved genitalia that look like the male version of female genitalia). Indeed, females with the more masculine version experience a lower number of traumatic mounts than those who do not.

Is homosexual traumatic mating necessarily a bad thing? Consider the case of the pirate bug (*Xylocoris* species), where it has been observed that sperm collected from a male partner may actually transport itself through the haemolymph of the receiving male to his testes. Once in the testes it is likely to become part of an ejaculate destined for a female target, essentially providing the possibility for a male to successfully fertilize a female indirectly through another male. This scenario has yet to be conclusively demonstrated, although a similar story has been described for homosexual interactions between male seed beetles. If it's possible to fertilize a female by first fertilizing a male, then it could be evolutionarily advantageous to do so; however, it does seem to be a rather roundabout way of increasing one's biological success.

One thing we have yet to discuss is the fact that males may utilize the power of traumatic intromission to remove other males from the pool of sperm competition. In earlier sections of this book we learned about the diverse ways in which males act to limit each other's reproductive success. Why should traumatic stabbing be classified as any different? If it *is* possible for a male to discern the sex of a potential male partner, and he performs traumatic intromission anyway, it could perhaps be assumed that he is attempting to decrease the reproductive success of his homosexual partner by sexually wounding him and therefore tying up his energies in attempting to heal. In this way, more dominant males can copulate their competitors into submission, rendering them less able to achieve reproductive success with females.

Give the penis a bone

The penis bone, or baculum, is an enigmatic structure possessed by males of various species across all mammalian orders, the carnivores, bats, rodents, insectivores and primates. Penis bones are heterotopic, which means that they are derived from bone tissue that has been formed outside of the skeleton. All bacula are synthesized from connective tissue located at the distal (tip) end of the penis, just above the urethra; however, their morphologies between organisms are extremely diverse. Due to their vastly variable appearances, penis bones are known for their large role in taxonomy and species identification. Some of them are smooth, others have finger-like extensions or hooks, some are toothed, and some are spiny. They can be either large or small for body size – the largest belonging to that of the walrus (up to 60 centimetres!). Humans don't have penis bones, and there is an overall trend towards their reduced size in other great apes. One of the most interesting things about penis bones is that scientists cannot come to a robust consensus with respect to their exact purpose. Their functions, sizes and morphologies vary markedly within and between mammalian orders – so a specific task or pattern observed in one organism may be completely absent in another.

One of the main reasons for the lack of consensus is their sheer diversity – even within individuals of the same species.

A major school of thought when it comes to the function of the penis bone is structural support. Seeing as they are ossified bones, there is no question that they play a supportive role in achieving and maintaining erections. Indeed, in several primate species with prolonged copulation durations, males have longer penis bones. Sex can be violent and potentially harmful to a penis, especially the kind that takes place between land-breeding pinnipeds (seals, sea lions and walruses), and while penis bones can be fractured, they can also provide a degree of protection to the otherwise vulnerable phallus. Another potential function of the baculum is to protect the urethra from compression, although there is at present little support for this idea. Perhaps we should shift this discussion to one of the more supported views of baculum function: to keep the ladies happy.

It sounds like something out of an erotic novel, but there are biologically relevant aspects to the nature of female titillation during copulation. The morphology of the baculum certainly affects the degree of stimulation received by a female, and in some extreme cases they have evolved their own bacula-like structures to fit into those of the males. For example, in ground squirrels (in the family Sciuridae), females have evolved vaginal folds that interdigitate with the protruding teeth on the male's baculum. Called a baubellum, the female version of a penis bone is also well developed in a few primates, such as ring-tailed lemurs where it reaches half of the size of the male's baculum, although little is understood about its form and function. (Protection? Stimulation?)

When would it make sense for a male to want to stimulate his female sexual partner into submission? It's possible that females judge males based on characteristics of their bacula, especially if certain ones feel better than others. Indeed, the possibility exists that superior stimulation may function to reduce the probability of females re-mating, and there is some evidence of this in several primate species. Specifically, keratinized penile spines (another aspect of the complex genital systems of primates) are hypothesized to increase copulatory stimulation, and perhaps play a role in decreasing a female's window of sexual receptivity. The flip side of this argument is that a certain level of damage to a female's reproductive tract, inflicted through penile spines or other baculum characteristics, will also limit her immediate future reproduction while her body heals. If a male does a small amount of physical damage to a female during copulation it could actually benefit him in terms of biological fitness.

There are other ways in which male genitalia may allow a particular male more exclusive access to a female. First, by maintaining his erection (through the structural support afforded by the baculum) for longer periods, a male may increase the chances that his sperm will be the winning sire by keeping his penis inside her body subsequent to ejaculation. Some more elaborate bacula may also function in removing sperm of previous suitors from a female's reproductive tract, or inducing ovulation.

Outside of the possible hypotheses for the evolution of the baculum in the first place, there are many ecological factors that contribute to their diverse structural and functional tasks. In some cases the social system may affect bacular morphology among males. Sexually mature male bank voles (*Myodesglareolus*) are either socially dominant or not. Dominant males have a significantly wider baculum than do submissive ones, suggesting that these traits evolved in conjunction with each other. In polecats (*Mustela putorius*), baculum size is hypothesized to be a reliable indicator of male quality because it continues to grow throughout their lives. Males with more energy to contribute to bacular growth are apt to be more biologically fit. In a more general sense, species with multi-mate mating systems have more well-developed bacula than do species that are socially or serially monogamous, indicating that there are cases where sperm competition and cryptic female choice have played a role in the evolution of the penis bone. This idea was experimentally tested using house mice (*Mus musculus*). In an elaborate laboratory setup, four separate populations were propagated over twenty-seven generations with specific mating strategies: multi-males mating with a single female or a strictly monogamous scenario. The results showed unequivocally that the mating system can directly affect baculum morphology, as males in the multi-mating scenario evolved thicker bacula than ones that were monogamous.

In addition to the type of mating system employed, there are certain environmental characteristics that may play a role in the evolution of penis bones. For example, many organisms in high latitude (Northern) climates occupy substantial home ranges, and roam across large areas in search of resources and mates. This can be quite a task, especially when finding a mate means traversing through ice and snow. Northern mammals tend to be solitary, and are associated with 'bet-hedging' life history strategies, and males often have large, long penis bones. Ovulation is frequently induced by copulation due to the low encounter rates with the opposite sex – and a long penis bone may function in this way, through stimulation of the female's reproductive tract. For many of these organisms, the hostile environmental conditions like large amounts of ice and snow can have negative effects on the frequency of mate encounter. So having a great big bony member to ensure that your mate ovulates when you have sex with her is decidedly advantageous.

Girls with boy bits

Did you know that human females have penises? The clitoris is assumed to be the female's version of male genitalia, a vestigial structure that remains in the female body because of its immense importance in males (see 'The Big O'). The mere evolutionary byproduct that is the clitoris is a source of pleasure for orgasmic females everywhere. Despite this evolutionary homology (i.e. the structures are from a common origin but have since diverged), human genitalia between the sexes are morphologically distinct. There are girl bits, and boy bits, and they are easily identifiable. This isn't the case for all mammalian females. In some truly astonishing twists of evolutionary oddness, ladies in several mammalian orders have evolved external genitalia resembling that of their male counterparts. The penile-clitoris is the name given to these structures, which are essentially hypertrophied clitorises that resemble the phallus. Pseudopenises are most well known for their presence in elephants (family Elephantidae) and spotted hyenas (*Crocuta crocuta*), and they make it nearly impossible to differentiate between the sexes using mere external morphology. In elephants, males and females have an identical anogenital distance, which means that the amount of physical space between the anus and the penis (or penile-clitoris) is exactly the same. Moreover, males have no scrotum, which is usually a handy way to determine maleness in mammals that have retractable penises. However, social context in elephants does provide fairly reliable information about the sex of individuals. Adult female elephants are never without a group, whereas adult males are never part of one (unless a male is copulating with a female). Females live together with other females and offspring, and when males reach sexual maturity they are evicted and forced to live as solitary individuals or as part of small bachelor pods. When a female is in oestrus (which is rare, approximately two to three days every four years), an adult male is permitted entry into the group for copulation purposes only. Which male, you ask? A delightful physiological and behavioural process called musth determines the identity of the lucky suitor.

Males that are 'in musth' experience a five-fold increase in circulating testosterone levels and are extremely aggressive towards other males. If a different (non-musth) male attempts to copulate with an oestrus female, chances are he will be killed by the one in musth, who will be making strange vocalizations and emitting discharges from sweat glands and his penis.

As you can imagine the astonishing genital structure of female elephants means that the act of copulation is not simple. Actually, it is physically impossible for a male to have sex with a female unless she is completely willing, and this is because she must first retract her penile-clitoris into her own body (think about turning a sock inside out). Only once it is safely tucked inside the female's body can the male insert his penis into her vagina. To put it another way: it's physically impossible to rape a female elephant. Females will only ever allow males to have sex with them during the brief period of ovulation, and each act of copulation lasts less than one minute (though there will be several throughout the day). So as if it isn't bad enough lacking the capacity to copulate with a female whenever he wants, when a male elephant finally does get to copulate it lasts for an extremely short period of time. When it comes to sex, it's not much fun to be an elephant.

As I mentioned earlier, the other lady that's most famous for her boy bits is the spotted hyena (*Crocuta crocuta*). These females have extremely complex penile-clitorises, and their external labia are fused together to form a pesudoscrotum. So if you thought you'd be able to sex a hyena by looking for testicles, you'd be wrong. The genital structures of female spotted hyenas win them the distinction of being the only mammalian females to copulate, urinate and give birth through the penile-like canal. It's even possible for females to achieve erections. There are some slight differences between the genital structures of males and females, namely that the male penis is longer and thinner with an angular head. These characteristics make it easier for him to insert it into the female's genital opening (once she's retracted her penile-clitoris).

Although the external genitalia are morphologically similar to those of males, the internal urogenital system of females is more traditionally feminine. She gestates offspring in her uterus; however, as I mentioned above she also gives birth to 1.0–1.5 kg babies through the penile-clitoris. The birth canal is approximately 2.5 cm in diameter, and females often experience severe tearing. The rate of female mortality during childbirth is abnormally high for first-time mothers. The news isn't all that much better for the babies, who often separate from the placenta, get caught in the birth canal and die from anoxia. What are the evolutionary origins of such structures? In what possible circumstances would it be advantageous for females to evolve such highly masculinized genitalia? There are other mammalian females that possess hypertrophied and masculinized genitalia, such as moles and lemurs, but none possess the remarkable structures of the spotted hyena.

Of the four species in the family Hyaenidae – spotted (*Crocuta crocuta*), striped (*Parahyaena brunnea*) and brown (*Hyaena hyaena*) hyenas, and aardwolves (*Proteles cristata*) – only the spotted hyenas have acquired the penile-clitoris.

There are a few schools of thought with respect to the evolutionary origins of this remarkable structure. Hormones called androgens have always been present in the systems of female spotted hyenas, and these substances are associated with dominance and social status. In addition, testosterone levels in spotted hyena females are as high as those in males, unlike female striped or brown hyenas (who are also not socially dominant to their male counterparts). It's thought that high levels of testosterone are transferred to male and female offspring through the placenta, exposing both sexes to a high level of maleness during gestation. These steroids in adult females make them extremely aggressive, much more so than their male counterparts, despite being of similar physical size. A clear advantage for females to be dominant is illustrated in the feeding system of spotted hyenas, which converge on freshly killed prey and eat as a group. This is quite unlike the solitary foraging tactics employed by striped and brown hyenas. Over twenty individuals may be vying for a piece of a kill, so competition is high. It's been noted that a group of spotted hyenas can transform a wildebeest to mere stains on the grass in a matter of minutes. High-ranking females are the first to indulge in a kill, along with their offspring. Low-ranking females and *their* offspring are next on the list, finally followed by adult males. Without question, females are socially dominant to males, so there are clear ecological advantages to having high levels of androgens and testosterone, and the 'nonadaptive' hypothesis suggests that the extremely masculinized genitalia of these ladies is an evolutionary by-product resulting from possession of these steroids.

There is some merit to this hypothesis, but thinking back to giving birth through a penis (a *massive* biological trade-off) makes me wonder if there is another hypothesis that explains the existence of the penile-clitoris.

An alternative school of thought behind the existence of the highly masculinized genitalia of the female spotted hyena is that the structure evolved at least partly in a social context. As I've stated above, high-ranking females are extremely aggressive towards both males and lower-ranking females. They realize a higher level of nutrition and reproductive success than subordinates, and have even been observed to attack lower-ranking females and their offspring. It would seem then, that existence of a ritualistic behaviour that reiterates social status may be advantageous for high- and low-ranking females, to maintain dominance in the former and to reiterate submissiveness in the latter. When females greet each other, they stand in a parallel position, facing in opposite directions. They then lift their hind legs and display their fully erect penises to each other. They will lick and smell each other's penises, but there is a suite of behaviours that are performed only by the more submissive individual. Asymmetries in these greeting displays are greatest between highest- and lowest-ranking females, suggesting that this could represent some kind of mutual communication about social status. It would make sense for a low-ranking female to engage in whatever submissive postures she can because this position leaves her most delicate bits completely vulnerable. Although this 'submission signal' hypothesis explains the existence of these gestures, I'm doubtful that it's enough to drive the evolution of the female penis. Remember: these females are giving birth through a 2.5-centimetre canal!

A third hypothesis for the evolution of the penile-clitoris has to do with the notion of sexual monomorphism (both sexes look the same). There are widespread species in which males and females closely resemble each other for different ecological reasons. It's thought that sexual monomorphism in spotted hyenas could be advantageous for a few reasons. First, there is intense aggression between newborn hyenas. These are not your average 'cute baby animals'. Newborn hyenas are located in underground dens that are not easily reached by their mother (or any other adult). They have functional incisors, large canines and muscular necks and jaws. In addition, they are extremely aggressive, often attacking anything that moves. Interestingly, female newborns experience a greater level of siblicide than males. Why would this be the case? In hyena societies, males are essentially the lowest rank on the totem pole, whereas there is intense aggression between females of various rank (see above). A male newborn is therefore a much lesser threat to any other newborn in the litter than is a female. Furthermore, high-ranking females have been observed to kill the female offspring of lower-ranking females, again substantiating a biological desire to keep the pool of competitors low. When spotted hyena pups are born, they are virtually identical. They can only be sexed through molecular means; male and female genitalia are indistinguishable. The physical resemblance between males and females decreases from infancy onwards – in fact it's easy to sex a post-reproductive adult female through her torn penile-clitoris and pulled nipples. This suggests that pressure for females to appear as males is greatest at early stages of a female's life. This idea for the evolution of the penile-clitoris makes the most sense to me. It's a structure that comes with both immense benefits (social dominance, sexual dominance) and immense costs (I'll say it one more time: they have to give birth through a penis), and I feel like it is just unlikely to have evolved as a byproduct of some other structure or behaviour. The sexual mimicry hypothesis, where females have evolved genitalia that is indistinguishable from males – especially at infancy where they are most likely to experience higher levels of siblicide or infanticide based solely on their female identity – seems the most plausible candidate.

Transvestites

Outside of the extreme examples of females with male-like genitalia, there is no shortage of cross-dressing in the animal kingdom. Girls posing as boys, and boys posing as girls. At first thought transvestitism may seem counterintuitive, but ecological complexities sometimes mean that cross-dressing is not only advantageous, it's critical in maximizing biological fitness. One of the best-studied examples of animal transvestites comes from the damselfly (*Ischnura* species) world, where it is exhibited by over a hundred different species. Female damselflies come in two very distinct morphologies: gynomorphs are the 'girlie girls', the true females of the population. These ladies range in colour from orange to brown, and are very distinct from males. Andromorph females are bright blue and green, and bear a striking resemblance to males. Now, why would a female want to appear as a male? As with many invertebrates, female damselflies have a sperm storage organ (spermatheca) and they can obtain all the sperm that they need for their entire reproductive lifespan in one or two copulations. The sexual behaviours between males and females are therefore disturbingly different. Males copulate daily, and almost all these copulations are forced upon unwilling females.

The 'male mimicry hypothesis' posits that the andromorph female form evolved in response to the extremely high level of sexual harassment in many of these species. Indeed, gynomorph females experience a higher level of harassment than andromorphs, and this could have direct negative impacts on their health and biological fitness. So you may be thinking, why not move towards a monomorphic (one physical appearance only) morphology overall? Why should the gynomorph morphology be evolutionarily sustained if these ladies receive a higher level of harassment and a lower level of fitness? As with any biological system, there are more factors to be considered.

In low-density populations where harassment levels are low, gynomorph females do just fine. It's in the high-density populations that we see a substantial level of harassment. In addition, gynomorph females are more cryptically coloured against their natural grass backgrounds, meaning that they enjoy a lower level of detection by predators. Male damselflies primarily use their vision to find females, so cryptically coloured gynomorph females may, in some low-density scenarios, be more difficult to find. Gynomorph females undergo a drastic colour-change during development – their dorsal side goes from red (immature) to dull brown or orange (mature). Andromorph females do not experience any colour change of this nature, nor do males. This ontogenetic (developmental) change in coloration seems to allow males to identify when gynomorphs are immature. They receive a significantly lower level of harassment as juveniles than do andromorph females, whose colour remains constant throughout their lives. Overall, it seems that in varying ecological circumstances there are advantages to being either gynomorph or andromorph, which explains the persistence of both morphologies in damselfly populations.

There are other cases where transvestite females may be disadvantaged by appearing as males. It's fairly common in sexually dimorphic creatures (with highly ornamented males) for females to display rudimentary versions of the males' sexual ornaments. The masculinized traits are a byproduct of the genetic control of sexual ornaments in both sexes. Think about it this way: facial hair is a sexually selected trait in human males, yet there are some females that bear facial hair due to their genes – albeit to a lesser extent. These 'bearded ladies' may suffer fitness consequences for being masculinized. For example, Eastern fence lizard (*Sceloporus undulatus*) males have distinctive colour badges on their ventral sides. These bright blue patches outlined in black are a striking feature of the throat and abdomen. Some female fence lizards also display these colourful badges, but to a much lesser extent than their male counterparts. The patches serve major ecological functions in the lizards: they are the primary identifier of sex to potential mates. When males' patches are covered up, other males actively court them, and when females receive painted patches they are treated aggressively by males. Levels of plasma testosterone mediate the formation of the badges, so it follows that aggression would be a common characteristic in males that have large ones. Female lizards that bear rudimentary patches are less preferred by males. They are approached less often despite the fact that males can utilize chemical cues to determine their sexual identity. These 'bearded ladies' lay a smaller number of eggs than their non-bearded counterparts, and their eggs are laid much later in the season, perhaps reflecting the reluctance of males to mate with them. Here, there are clear fitness consequences for appearing as a male. It's possible that there is a social function for being more masculine, perhaps in achieving social dominance over other females; however, it seems most likely that in this case being a cross-dresser is a negative trait under genetic control.

Although there are many documented cases of females taking on male characteristics, the vast majority of cross-dressing in the animal kingdom consists of males dressing like females. Let's face it, a common social scenario is one of large, dominant males that hold territories and keep harems of females. An unfortunate reality that goes along with this scheme is that males have to be small before they are big. Small males often take on a female-like morphology in order to avoid aggression from territorial males that could easily dominate them. Small (yet still sexually mature) male Galapagos iguanas (*Amblyrhynchus cristatus*) utilize their female-like appearance to sneak on to the territory of a larger male to seek out copulations. These smaller males masturbate, then store the ejaculate in small pouches near their penis to reduce the time it takes for successful copulation. Another well-documented example is that of giant cuttlefish (*Sepia apama*), where small (yet sexually mature) males will utilize the immense capabilities of their system of chromatophores (pigments on their outer integument that are intricately linked to both nerve and muscle cells and allow for diverse patterns of signals and camouflage) to appear as a female and approach sexually mature females even when a large dominant male is nearby. Nearly half of these cross-dressing mating attempts are successful. This kind of small-male transvestite strategy is common in all kinds of birds, fish and lizards. Males start off small and feminine, and eventually grow out of these characteristics and embrace maleness once they are large enough to defend themselves. However, there are a limited number of organisms where males live in permanent, genetically determined, transvestite forms.

Such a lifestyle has been documented for two bird species, the lekking ruff (*Philomachus pugnax*) and the marsh harrier (*Circus aeruginosus*). It's thought that these male polymorphisms originally evolved as a means of reducing aggression between males, an extension of the juvenile female-like forms I described above. In the former, she-males are slightly larger than true females, and they have testes approximately 2.5 times the size of normal males – suggestive of a sneaking strategy for successful reproduction. Having a lot of sperm at your disposal is handy when you're shooting it off as often as you possibly can. In the marsh harrier, she-males make up approximately 40 per cent of the population, a substantial portion. Here, cross-dressing appears to reflect a kind of 'non-aggression' pact between males. When presented with either male or female decoys, the two male morphotypes react very differently. Traditional males are more aggressive towards other traditional males than they are towards she-males. Interestingly, she-males are less aggressive towards traditional males and more aggressive to traditional females. So not only do they resemble females, these males behave like females, acting aggressively in order to defend breeding resources.

Multiple male morphotypes in which one appears as a female have been observed in a few other species including a marine isopod, a fish and a lizard. In all cases it's important to keep in mind that these different male morphologies are evolutionarily stable. They would have failed to evolve if the reproductive fitness of any one of them was lower than that of the traditional male morphotypes, so it's worth examining aspects of their ecology with that assumption in mind. Consider the side-blotched lizard (*Uta* species), which has three distinctive male morphologies as expressed by colourful throat patches. Males with bright orange throats are the most aggressive and defend large territories; those with dark blue throats are less aggressive and defend smaller territories, and they engage in mate-guarding. The third kind of male is the transvestite, with yellow stripes on his throat patch that resemble those of females. As I mentioned earlier, all three of these male types must experience relative equality when it comes to reproductive success, and indeed they do, but the most abundant males vary in cycles through time in a fairly predictable way.

Almost akin to a game of rock-paper-scissors, the abundance of aggressive orange-throated males will eventually give way to an abundance of yellow-striped transvestite males through the successful sneaking strategy of the latter. The blue-throated males will then become more dominant because their mate-guarding strategy undermines the sneaking behaviour of the yellow stripes. However, the orange throats are able to take over from the blue throats by being more aggressive and dominant. This cycling of male-morphologies occurs over a time frame of approximately five years.

These examples suggest that the commonality of males cross-dressing as females exists primarily to decrease tension and aggression between males. For the most part, this is the main reason behind transvestitism in the animal kingdom, although there are cases where males appear incognito as females for other purposes. Take the red-sided garter snakes of central Canada (*Thamnophis sirtalis parietalis*), who have made more than one appearance in this book for their bizarre sexual behaviours. As we have seen, these organisms hibernate in underground dens by the tens of thousands, and upon waking in the spring engage in mass orgies that last for several days at a time. Males who wake early gain a considerable advantage over males that wake later (in terms of the number of copulations they can attain); but it's not always advantageous to be the first man at the party. This is because later-waking males secrete female pheromones (an indirect form of cross-dressing), which attracts the early-waking males to them. The 'fooled' males attempt to copulate with the 'fooling' males, and in doing so, provide some much-needed body heat that helps the fooling male to warm up more quickly and begin his own copulatory adventures.

A similar scenario is found in the grain-breeding parasitoid wasp (*Lariophagus distinguendus*), where adult females oviposit individual eggs in clusters on stored grain. Each individual develops within its tiny case, but many cases are situated in clusters. Developing females produce pheromonal cues from within their egg-cases that allow adult males to find them by waiting for them to emerge and inseminating them before any other male gets the chance. In a similar fashion to the red-sided garter snakes, male parasitoids that emerge late relative to other males produce a pheromonal signal that mimics that of the females. In this way they waste the time of their direct competition by having them wait around for a female that's actually a male. To put it another way, the late-emerging males are mimicking females in order to stall the reproductive efforts of their direct competitors.

I'll end this discussion with another spectacular example of cross-dressing from the invertebrate world. Utilization of the complex system of chromatophores for disguise was described for giant cuttlefish above. Another species, the mourning cuttlefish (*Sepia plangon*), also undertakes a form of cross-dressing, although not in the small versus large male context. Populations of mourning cuttlefish are male biased, leading to intense competition for female mates. Males will engage in mate-guarding and displacement of rivals, even attempt to disrupt the copulation attempts of others. In an extreme stroke of evolutionary genius, males will display courtship patterns to females on one side of their bodies, while concurrently displaying deceptive female patterns to rival males located on the other side. In other words, simultaneous dual-gender signalling, which demonstrates an astonishing level of cryptic control. This kind of display also demonstrates an extremely high level of cognitive ability, seeing as it only ever works when a single rival male is present. In social contexts with more than one male or female, this dual-display strategy would not be successful because the disguise would be obvious to at least one of the parties present.

STIs

In today's world of sexual openness and empowerment, it's completely normal to check with partners (preferably beforehand) about their sexual history and their latest round of testing for any kind of sexually transmitted disease. If you're human, STIs are pretty easy to avoid through careful partner selection and/or appropriate use of sexual protection. The same kind of easy avoidance of STIs is not available to other members of the animal kingdom, and unfortunately sexually transmitted diseases are common across all animal phyla. The study of STIs in animals can be both daunting and extremely difficult since in many cases even observing copulation can be nigh on impossible, much less the observation of tiny microorganisms that come along with it. Along with the viral and bacterial infections that plague human populations, animals are subject to sexually transmitted fungi, protozoans, nematodes, helminths (worms) and arthropods. In fact, it's been said that no animal group that experiences internal fertilization can possibly be exempt from STIs.

Organisms from tiny fruit flies to gigantic whales may fall victim to infections by way of their sexual conquests. Indeed, Antarctic whalers from British whaling factory ships in the late

1940s believed that sperm whales (*Physeter macrocephalus*) carried venereal disease. One can assume that they believed this due to the fact that some kind of genital perturbation was consistently seen on specimens brought in for meat consumption. A study conducted in the late 1980s confirms the (previously ridiculed) claims of the whalers. Approximately 10 per cent of adult male bulls from a large sample caught off the western coast of Iceland in the 1980s carried lesions of genital papillomavirus on the ventral, lateral and dorsal surfaces of their penises. In addition to the active lesions, the tissues surrounding them showed clear signs of scarring and depigmentation. Follow-up data is not available. However, one can safely assume that even the largest animals on the planet are not immune to sexual infections from the very smallest.

One of the most extensively studied cases of a sexually transmitted disease is that of *Chlamydia pecorum* on indigenous populations of koalas (*Phascolarctos cinereus*) in Australia. Koalas are an endangered marsupial species, and certainly gain a lot of media attention by virtue of the fact that they are cute, cuddly and a patriotic symbol of the land down under. Unfortunately, they are also the poster children for chlamydia. A variety of approaches, surveys and analyses all suggest that chlamydial infections (both *C. pecorum* and *C. pneumonia*) are present in virtually all Australia's wild and captive populations. The disease is usually manifest in one of two ways. Infections can occur on the mucosal surfaces of the eyes, leading to extreme inflammation, discharge and ultimately blindness. Alternatively, infections present in the reproductive tracts of males and females cause brown urine and wetness of the fur around the rump and tail (often called 'wet-bottom' or 'dirty tail' disease). Individuals with severe symptoms have difficulty urinating and copulating, and the worst genital-tract infections can lead to infertility. So why do koalas have chlamydia? I suppose that's a fair question, but it would be naive of us to think that this is the *only* animal to experience a massive population decline due to an STI (see the many other examples described below). The fact is, koalas are charismatic and cute, and people like them a lot. This is likely why we care about them enough to know that they are suffering from such a nasty epidemic. Koalas have chlamydia because bacterial infections happen. All the time.

Early natural history records show that lesions indicative of chlamydial presence were observed in koala populations in the late 1800s. These organisms are fairly sessile (i.e. they don't move around much on their own), although humans have had a hand in moving them to various new habitats within Australia and are therefore perhaps partly responsible for the spread of the disease. Transmission occurs via sexual contact between males and females (horizontal spread) and from mothers to their joeys (vertical spread). Koalas have a reproductive structure known as a cloacum, which is the term for an opening that is for both sexual and excretory purposes. As a precursory move to commencing their diet of eucalyptus leaves, young koalas eat a substance called 'pap' from their mother's cloacum. Pap is essentially comprised of pre-digested eucalyptus leaves and a full suite of digestive bacteria that joeys need to acquire for their own digestive tracts. By eating this 'faecal-leaf-bacterial' mixture from the mother's cloacum, baby koalas are also likely to orally obtain the *Chlamydia* virus.

Speaking of cloacas, amphibians, birds, reptiles and mono-tremes (egg-laying mammals) possess these structures. Let's take a few minutes and think about the fact that said openings are a location for sexual activity *and* excretion. In other words, it's a direct, unblocked opening to the digestive and urinary systems – which are home to an immense number of microorganisms (both pathogenic and not). All organisms have a massive and di-verse community of bacteria, fungi and viruses in the digestive and excretory systems; however, in those that have cloacas this microflora is automatically also associated with sex. We might therefore expect to see extensive amounts of sexually transmitted infections in organisms with cloacae. Indeed, we do.

An interesting phenomenon exists in the common lizard *Zootoca vivipara* – females are either monandrous (mating with only one male partner) or polyandrous (mating with several male partners). It is uncommon for multiple female sexual strategies to exist within a species, so this represents an excellent opportunity to examine the differences in sexually transmitted bacterial fauna from females with contrasting lifestyles. Not surprisingly, females that mate with several males have much more diverse bacterial communities in their cloacas, indicating the facility with which they are transferred. Cloacal bacterial communities (both pathogenic and not) provide biologists with an opportunity to examine their direct sexual transmission.

In several socially monogamous seabird species, examination of cloacal communities provides insight into just how sexually monogamous each partner is. Imagine being able to detect whether your partner is cheating based on a quick examination of their genital bacteria! Though sexual indiscretions are the rule rather than the norm in most socially monogamous species, seabirds are believed to be a specialized case of fidelity because of the high cost of raising young in harsh, unpredictable environments. Males are less likely to make a substantial parental investment in offspring that are not biologically theirs, and females cannot provide enough parental investment on their own – without her partner's help offspring cannot survive. This scenario is rare in the animal kingdom, it represents a case where it makes the most biological sense for a female to 'stand by her man' socially and sexually. Indeed, the STIs of monogamous seabirds are more common within mating pairs than between mating pairs of several species, indicating that extra-pair copulations (which could be identified by a more diverse infection fauna in female cloacas) are rare. In little auks (*Alle alle*), an Arctic seabird, the majority of extra-pair copulation attempts by males are unsuccessful due to a lack of interest by females, which suggests that females have no interest in jeopardizing their relationship with their partners (and coparents) by sneaking around behind their backs.

Despite the fact that STIs are common and diverse in birds, they do not have drastically negative effects on their hosts. Not all foreign infections are so innocuous and friendly. Most STIs are damaging, causing any number of dysfunctions. In addition to physical maladies, it may be more difficult for infected individuals to find a sexual partner if said partner is privy to the infection. After all, who could blame you for steering clear of a partner that you know has herpes or gonorrhoea? STIs can negatively affect the fitness benefits that a partner is able to provide, and the 'avoidance hypothesis' describes the situation where a potential partner (usually female) avoids a male due to a) his low fitness at not being able to avoid getting infected in the first place and b) his further decreased fitness due to the infection. As if this wasn't bad news enough for a male, the 'contagion indicator hypothesis' suggests that females will also avoid infected partners because they are unwilling to expose themselves and their potential offspring to said infections. If a man is bursting at the seams with disgusting lesions or fungal growths, it doesn't exactly make him a hot-ticket date. In fact, there is no shortage of horror stories about the horrific and violent deaths suffered by animals as a result of these sexually transmitted beasts.

Imagine the case of a poor cicada (*Magicicada* species) infested with the sexually transmitted fungus *Massospora*, so utterly afflicted that some of its abdominal segments have broken away, revealing a mass of infectious conidia (fungal spores). Despite overwhelming infestations, debilitated cicadas still fly around and attempt to mate. But this isn't just some desperate hope on their part that another cicada will look past their infectious mass. When considering the plight of animals affected by sexually transmitted infections, the biological needs of *three* organisms need to be taken into account: the male, the female, and the pathogen. And those pathogens want what any organism wants: the chance to survive and reproduce. Unfortunately, to do so, they need to spread, so it's in their best interest that their unlucky host keeps finding new partners. We should almost think of the pathogen as the evolutionary third wheel, because in many cases neither infected males nor females remain in control of their sexual desires. It is assumed that infected cicadas still try to mate because the fungus is driving them to do so for its own selfish purposes.

The sexual activity of male milkweed leaf beetles (*Labidomera clivicollis*) follows a more predictable trajectory when it comes to a sexually transmitted infection. The 'host compensation' hypothesis predicts that an individual carrying a burden of infection should increase their reproductive effort in the short term, possibly before their bodies succumb to the infection. Indeed, there are many examples where STIs cause a dramatic increase in sexual activity or receptiveness in individuals that have them. When the male milkweed beetles are infected with the mite *Chrysomelobia labidomerae*, they display a significantly increased vigour in front of females. They copulate more frequently and for longer periods than males without the mites. Overall it makes good biological sense for a sexually transmitted organism to up its own rate of transmission by changing the behaviour of its host to have more sex – although the specific physiological mechanisms by which this happens have yet to be discerned. In this particular case the mites increase their own rate of transmission by congregating on the male's body at the precise location where he attaches to the female. In one observation, a contact of less than two seconds resulted in a transfer of thirteen mites!

The fungus *Entomophthora muscae*, which infects common houseflies (*Musca domestic*), takes things a few steps further. It can be spread in a number of ways including through sexual reproduction, and it is extremely virulent, often causing death within one week of contraction. But even dead flies can serve a purpose. Fungal spores continue to develop in the abdomens of corpses, pumping them up to resemble large, gravid females. Healthy males are attracted to these dead bodies and they actively copulate with them, helping to spread the fungus. Interestingly, male flies are not attracted to uninfected cadavers, but when abdomens of dead, infected flies are glued to dead, uninfected individuals, males will readily copulate with them. We can therefore assume that the highly virulent fungus is having effects on the reproductive behaviours of male flies.

Host/parasite interactions need not involve such intense manipulation. The relationship between the parasitic mite *Kennethiella trisetosa* and the wasp *Ancistocerus antilope* can on some level be described as mutualistic. Adult female wasps lay their eggs in small cavities within larger nests. Into each small hole she deposits an egg plus paralyzed caterpillars for the developing offspring to eat. Generally, since adult females are infected with mites, a few of these are inadvertently dropped off as well. Females repeat this action many times and essentially create a linear set of developing caverns. Once the wasp larva has hatched, it eats the caterpillars and continues to develop to a quiescent prepupal stage. It's during this 'quiet' phase that the mites infest the developing wasp's body, feed in its haemolymph (body fluid) and copulate. Although there is certainly an energetic cost to the fact that the mites feed on the wasp, it appears negligible. Prior to chewing through the partition and escaping to the outside world as mature adults, female wasps kill all the mites that have made their way to her acarinaria (the region where the mites carry out their adult lives), and emerge from their nesting enclosures parasite-free. Males, however, do not kill the mites that exist on their bodies, emerging as healthy (yet infested) adults. Males transfer mites back to females through sexual reproduction, repeating the cycle. So why don't male wasps kill mites? Recalling that when it comes to sexual reproduction, the needs of males and females are not necessarily the same, it may make biological sense for a male to 'want' his female partner to be burdened with a sexual parasite because it may lessen her ability to copulate with additional males. That's right – males may gain a reproductive advantage by transferring both genetic material and sexual parasites to their female partners if the additional parasitic 'gift' reduces her chances of remating. Additionally, the wasps have a haplodiploid genetic system, meaning that males do not contribute genetic material to sons, only daughters. Since juvenile female wasps are able to kill the mites in their developmental chambers and adult males produce only daughters, there may be no biological reason for them to 'bother' ridding themselves of parasites. Although it has yet to be confirmed, there is speculation that the presence of the mites in the incubation chambers may have some positive effects on the developing embryos, which would render this relationship as one of mutualism rather than parasitism. In fact, having a sexually transmitted infection can sometimes render individuals more biologically successful than they would have been otherwise.

Consider the scenario of fruit flies (*Drosophila* species) infected with the (somewhat) sexually transmitted sigma virus. I say somewhat because the sigma virus is transferred directly through the sperm to the offspring. It's an intracellular virus that therefore does not infect the female sexual partner, but is present in the offspring. Males who carry the sigma virus sire more offspring compared to males that do not – despite no differences in number of courtships or copulations in an experimental setting. It seems, therefore, that females mating with infected males have an advantage in terms of offspring number, and they may additionally benefit by having infected sons with increased reproductive success. Biologists speculate that female fruit flies may increase their reproductive output in response to an infected partner, or they may bias fertilization in favour of infected males through post-copulatory sperm selection. It remains unclear as to whether there are actual reproductive benefits to the virus, or whether it is the virus causing the altered behaviour in its hosts. Fascinating as it is, this example is unusual and does not hold when females are infected with the sigma virus. Nevertheless, it's a delightful example of the power of evolution in the context of sexually transmitted infections.

The key requirement for the 'success' of any parasitic organism is that it has to walk a fine line between infecting its host(s) to a great enough extent to keep its own populations alive and well, while making sure it isn't so virulent that the host population dies out. Without a home, the infection dies as well. Populations of ladybird beetles (*Adalia bipunctata*) are subject to intense crashes and local extinctions, but there is more to these population crashes than a simple sexually transmitted infection. These organisms are already host to a bacterial infection that is inherited by sons and daughters (i.e. *not* an STD) from their mothers. This bacterium is called *Spiroplasma* and it is fairly widespread in insect populations worldwide. Affectionately termed 'male-killing bacteria', *Spiroplasma* lives up to its name; killing 75 per cent of males and creating skewed sex ratios in populations where it exists (there are no ill effects on females). Now, the male-killing bacteria are simply a fact of life for many insect populations. However, while not sexually transmitted themselves, their presence wreaks havoc on the progression of sexually transmitted infections in these populations because of the extremely skewed sex ratios that they create. For example, the sexually transmitted mite *Coccipolipus hippodamiae* imposes costs on its host's fertility and fecundity, and ultimately causes mortality. It does not affect mating rate (as many sexually transmitted infections do), but due to the extreme bias in sex ratio in ladybird populations affected by male-killing bacteria, males mate approximately four times more than females. This essentially makes the STI epidemic extremely male-biased, and can often lead to population crashes once all the males in a population have died. As if the presence of male-killing bacteria wasn't enough.

Unfortunately for the females of the world, the incidence of an STI epidemic being male-biased is fairly rare. STIs are more often a female problem because not all males in a population get to copulate, whereas all females generally do. This sets the scene for a prevalence of female bias in STI epidemics, and a trend that has been observed in the promiscuous invertebrates and vertebrates alike. The conclusion when it comes to infections is that they are ubiquitous. Microorganisms exist within every system in an animal's body including the reproductive bits, meaning that humans are both unique and fortunate to have the luxury of a conversation about these things prior to jumping into bed.

The amazing topics we've covered give me (and hopefully you) a new respect for the tangled web that is animal sexuality. The notion that sex is simple, that inserting the penis into the vagina is all that needs to happen for the successful propagation of genetic lineages is so naively false. I hope that, if nothing else, this section has given you a new appreciation for why sex is the most critical and important thing that any animal can do.

SECTION THREE

THE AFTERMATH

Although it seems like a minor miracle to have made it successfully through steps one and two of sexual activity, the fact of the matter is that most creatures do it with remarkable ease. Despite the hardships suffered along the way, the creation of offspring is the inevitable outcome of successful sex, regardless of whether that sex was pleasurable or not. The next hurdle for any animal is to ensure that its biological spawn live long enough to reach sexual maturity and keep the cycle going. There are, of course, a myriad of strategies for doing so. Interestingly, humans provide much more parental care than is biologically required. If we were akin to our primate cousins we'd shove our babes out of the nest once they hit sexual maturity, or even sooner. Despite our close relatedness we display pretty massive differences in our parenting techniques. Indeed, as with the tremendous variation in successful strategies for sex across the animal kingdom, there exists an equal set of tactics for which one can be a successful parent. In this section I will examine some of the more interesting ways that organisms have evolved to cope with the demands of their direct descendants. As we have seen in previous sections, parenting in the animal kingdom is a far cry from lullabies and snuggly blankets.

Plastic parents

As we learned in the 'Plastic Partners' section, there are many aspects of an animal's environment and ecology that have effects on behaviour. When it comes to finding a mate, environmental stochasticity or predation pressure can add a myriad of factors to the mix that could easily affect the outcome of courtship attempts. The same is true for parenting. In a perfect world parents would be able to direct the bulk of their attention to provisioning their offspring without regard to intervening factors. Humans have it pretty good in this respect; in the Western world we can care for our children without worrying too much about the stormy conditions outside or predators lurking in the closet. The same cannot be said for parents in the animal kingdom, and they have developed some effective strategies to cope with varying conditions beyond their control.

High levels of predation pressure inevitably select for parental plasticity. In several nesting bird species, parents decrease the number of visits to the nest to provide food for their nestlings. This is likely to reduce their own visibility to predators but also to avoid disclosing the location of their nests. Being able to react to the presence of predation pressure is critical, but there are also costs entailed for the lower level of food provided to offspring. On the other hand, some parents will step up attentiveness in response to predation pressure, either by camouflaging their nests or by taking direct action against predators that are trying to reach their nestlings. Such behaviour is costly, and is expected only if parental investment in a brood is already high (i.e. a substantial amount of work has been done and chicks are almost ready to fledge). Due to the high cost of increased parental vigilance and attentiveness, extreme predation pressure is expected to result in a bet-hedging strategy where parents effectively stop provisioning offspring (leading to their ultimate deaths) in order to conserve energy for re-nesting when the level of predation lowers. It is a tough choice. Bird parents such as wrens (family Troglodytidae), nuthatches (genus *Sitta*) and chickadees (genus *Poecile*) increase investment in their chicks by increasing egg mass, clutch mass and feeding rate of nestlings when predation pressure is relaxed. It makes sense for parents to invest to a higher degree during times of safety, since a time of increased risk could always be around the corner. Parents not only step up investment during egg-laying, but also during post-hatching development, which allows for chicks to reach a fledging size earlier.

Song sparrow (*Melospiza melodia*) parents not only respond to the direct presence of predators, but also to their perceived risk of predation. In some elegant experiments that simulated predation pressure, researchers removed all predators (squirrels and blue jays) from study areas, and then played their calls in the presence of nesting parents. In this way, there was no actual possibility that predation could take place, but of course the unsuspecting parents would not have known this. Brood sizes for song sparrows with a high-perceived risk of predation are 40 per cent lower than those without, indicating that predation risk plays a key role in the population dynamics of songbirds. Such a trend is likely to be evident in many bird species, with large effects across ecosystems.

The threat of predation may not be consistent in all areas where a species resides, leading to flexible parenting strategies depending on the immediate environment. For example, mothers of the long-tailed skink (lizard) *Eutropis longicaudata* rarely exhibit any form of maternal care throughout their large range in Southeast Asia. Once eggs are laid they tend to be left on their own. However, in a single population located on Orchid Island in Taiwan, mothers do exhibit parental care. Here, they actively deter egg-eating snakes from entering their nests and consuming eggs. Transplant experiments, where snakes from one population are translocated to a different place, show that there is a great deal of behavioural plasticity in the level of parental care offered by females. Those originally from Orchid Island (where snake predation pressure is high and maternal care is high) drastically reduced their level of maternal care when transported to a new location, whereas females transferred to Orchid Island expressed maternal care for the eggs that they laid there. Interestingly, the translocation experiments exhibited these results only when snakes were transplanted as young juveniles. Those transplanted as adults retained their original mode of parenting, regardless of the predator scenario, which shows that flexible behaviour is genetically retained but processes learned during development can override it.

While predation pressure plays an influential role in the plasticity of parenting, other factors, like resource quality, are also important. When resources are of high quality and abundance, we would expect to see parents taking advantage of the bounty and having larger broods or larger offspring. In many cases, this is exactly what we observe. Dung beetle (*Onthophagus atripennis*) parents provision their future offspring by collecting dung and transporting it to a series of underground tunnels that they have excavated for that precise purpose. Once at the blind end of a tunnel, the dung will be shaped into a neat little 'brood-ball'. Mum will deposit an egg into a specific chamber on the top of the ball, and she will then seal it with more dung. These balls play an important role in the development of the beetle larvae: this is all the provisioning that the parents will ever provide. One would assume that the quality of the dung balls is critical to the successful development and emergence of dung beetle babies. Parents are capable of assessing the quality of dung and adjusting their investment accordingly. When they encounter a high-quality pile (such as that produced by fruit-eating monkeys), beetle parents will use it to produce a large number of small brood masses, which will ultimately translate to a large number of high-quality emerging offspring. Conversely, when the dung is of low quality (such as that produced by ruminant cows), parents will produce a lower number of offspring, each with a larger brood ball. Unfortunately, even with larger fecal-balls to eat, beetle larvae reared on low-quality dung are inevitably smaller at emergence than those provisioned with high-quality food.

In addition to the quality of the food resource, the level of paternal investment by male dung beetles varies with social context and morphology. Males in the genus *Onthophagus* have one of two distinct morphotypes: those larger than a critical body size develop a pair of disproportionately long horns, and those smaller than said critical size only develop rudimentary horns. There are costs and benefits to both morphologies for males; those with the large horns tend to take on a dominant role in reproduction, whereas those with small ones gain reproductive success by sneaking. Paternal investment also varies between the morphotypes. When no other males are present, males with large horns provide a large degree of parental care in the form of help with tunnel excavation and dung transport, while those without horns do not. However, when other males are present (i.e. when there is competition for mates), males will cease any paternal behaviour in favour of mate-guarding. For dung beetles, competition for breeding opportunities is particularly intense, dung patches rapidly lose quality (within twenty-four hours) and so they must be immediately exploited if they are to be useful. If a lone pair happens to find a rich resource, a greater level of investment into offspring can be realized, but when competitors are present it becomes more important for males to secure their female rather than to help her with offspring provisioning. In this way, the social context has an effect on the level of parental investment.

The density of potential competitors also has an effect on the paternal investment of male gladiator frogs (*Hyla faber*) in Southeast Brazil. These males are extremely territorial and violent towards one another, and any kind of paternal care is rare. Egg masses are extruded by females, fertilized by males and normally abandoned by both parents to develop (or not) into tadpoles. However, when frog densities are high, male gladiators engage in egg attendance, which decreases nest intrusion by other males. Males will guard the eggs that they fertilized the previous night, which improves their offspring's chances of successfully developing while at the same time reducing the male's chances of finding and mating with another female during this time. As with any change in parental investment, the costs and benefits of specific actions are taken into account in order to maximize biological fitness.

The tree frog *Dendropsophus ebraccatus* utilizes a strange form of parental plasticity that is almost unique in the animal kingdom. These parents can deposit their eggs in either terrestrial or aquatic habitats, which is extremely rare seeing as the adaptations for aquatic eggs are quite different than those for terrestrial ones.

In water, eggs tend to be specialized to extract oxygen, whereas on land eggs are specialized to retain moisture and avoid desiccation. So the eggs of these tree frogs clearly possess a complex mix of adaptations to both environments since they can withstand being dry or completely submerged. Although the aquatic egg condition is ancestral (making the terrestrial condition more evolutionarily recent), there are costs and benefits to laying eggs in each kind of environment. Selection for terrestrial eggs allows for a relaxation of predation pressure from aquatic organisms, and decreased constraints on oxygen uptake. In addition, the ability to deposit eggs terrestrially in trees that hang over fast-moving streams is particularly relevant to several species of amphibians, whose tadpoles can drop directly into the water body. Fast-moving waters are likely to wash egg masses away, but swimming tadpoles can control their own movements in running waters. The main drawbacks to terrestrial egg deposition are of course exposure to terrestrial predators and the threat of desiccation. It appears that the latter factor is critical to egg deposition in this species of tree frog. When there is no shade present (i.e. threat of desiccation is high), parents deposit their eggs in aquatic habitats. When there is abundant protection from sunlight, adults lay terrestrially. Parents can even vary their egg-laying behaviour (aquatic versus terrestrial) during the course of a single evening. This fine-tuned sophistication of egg-laying allows tree-frog parents to maximize hatching success of future offspring according to immediate environmental conditions.

Flexible parenting is clearly important in a number of scenarios (as outlined above), although the ability of the *Dendropsophus* tree frogs to vary their laying environment displays a level of plasticity that most organisms do not possess. What about species that can produce offspring in different ways? Although rare, there are three identified (so far) amphibian species where females can engage in either egg-laying (oviparity) or give birth to live young (viviparity). The lizard *Zootoca vivipara* is one species that demonstrates this capability, although viviparous and oviparous populations are almost always geographically separated. Viviparous females commonly have a larger body size, which makes sense considering that they will carry eggs with an increasing volume for a longer period of time than their oviparous counterparts. Oviparous females generally experience higher energy expenditure during vitellogenesis (yolk deposition) and the first phase of development prior to egg-laying. The shells of oviparous females are eight times thicker than those of viviparous females, which makes sense based on the fact that the latter part of development will happen within the egg outside of the female's body. The energy saved by viviparous females is spent during the later phases of embryonic development, where offspring require a greater amount of maternal investment. The fact that two modes of reproductive development occur in these species is perplexing. It could represent a late-stage step in a speciation event (i.e. they are well on their way to becoming separate species), as the transition from oviparity to viviparity has occurred multiple times in diverse vertebrates. Or perhaps we can look at this as an extreme case of parental plasticity. It would be interesting to investigate whether oviparous females are also capable of viviparous development; perhaps some biologists somewhere are currently trying to make that happen.

The fire salamander *Salamandra salamandra* is another organism with a perplexing level of parental plasticity. Females are generally oviviparous (not to be confused with oviparous), which means that somewhere between thirty and sixty larvae are produced and birthed at a developmental stage prior to metaphophosis (the transition from aquatic, free-swimming larvae to terrestrial adults). The only food source for these offspring is their yolk sac.

However, females are also capable of an alternative developmental mode: viviparity. Viviparous females produce between one and fifteen fully metamorphosed juveniles, who obtain nutrition from their yolks and from direct maternally derived sources. Fire salamander mothers who are oviparous versus viviparous are not geographically separated, but it's not clear whether females are capable of individually switching their developmental mode. It's thought that evolutionary changes to the timing of development are responsible for the shift to viviparity in this species. Such changes to developmental timing are called heterochronic shifts, and they are responsible for a wide variety of speciation events. It's important to remember that the process of evolution need not act exclusively on the adult form of any animal (or even at the level of the entire individual, as we have discussed earlier), that *developmental phases* are as likely to experience selection pressures that result in overall changes to adult morphologies. In the case of the fire salamander, it's thought that development of the cephalic and pharyngeal structures (the head parts that are involved in feeding) happens at an earlier time in viviparous individuals. The fact that some embryos are capable of feeding at an earlier stage than others allows for them to cannibalize many of their siblings in utero, which contributes to their continued development to metamorphosis whilst still in their mother's body. It's a wacky notion to wrap your mind around, but this kind of change in developmental timing could have easily facilitated the evolution of another reproductive mode (i.e. viviparity) in the fire salamander.

Go ask your dad

Having gone through four pregnancies, four childbirths and four phases of lactation, I can attest first-hand that female humans have a much larger job when it comes to offspring care than males do. As we know, the female investment in almost all animals is much higher than that of males based solely on the value of eggs versus sperm; however, in mammals we've been dealt a double whammy when it comes to infant care. Not only do females have to gestate embryos within their own bodies, they then become the sole providers of nutritious food for newborns in the form of milk. How many times did I wish that the father of my children was able to provide some nourishment for those offspring so I could have a break. Alas, the biology of the situation is such that this is an impossible dream for female mammals of many species. The fact that male investment is particularly low is generally attributable to a lack of certainty over their paternity, since genetic monogamy in the animal kingdom is rare (exceedingly so for mammals). In addition, since females are physiologically unable to desert a developing offspring, it naturally follows that males should always seek to maximize their own biological fitness by impregnating others rather than sticking around to help out where he cannot be much use anyway. So for the most part it's simple: females are stuck with child rearing and lactating. However, the beauty of biology is that there are always exceptions to every rule – and it's in those exceptions that I find the most joy and amazement.

You see, it *is* physiologically possible for males to lactate. In fact, male and female nipples in humans and other primates are virtually identical until sexual maturity and its associated hormonal changes. In addition, it is not actually the pregnancy and childbirth that facilitate the process of lactation in primates. Milk production is stimulated by hormonal changes, specifically surges in prolactin that occur in the body. Such increases can be stimulated in a female body that has not experienced a pregnancy, through artificial means or through suckling and nipple stimulation. Interestingly, male mammals also experience surges in prolactin both spontaneously and through manual nipple stimulation, which certainly piques my interest about the possibility that in some future world we may be able to apply some kind of prolactin patch to fathers of newborns and allow them the joy of feeding their own babies. In the present world, however, there appear to be at least a few male mammals that are actively lactating and providing this nutrition to their newborns.

Male Dayak fruit bats (*Dyacopterus spadiceus*) in Malaysia and masked flying foxes (*Pteropus personatus*) in New Guinea have mammary tissue akin to that of females. Although their nipples are smaller and contain less keratin than those of females, they have the ability to express milk to nourish young pups. These bat species are quite elusive and not much is known about their general biology and ecology, but this represents a truly fascinating prospect about the evolution of lactation in male mammals. In organisms with a high level of paternity certainty and genetic monogamy, it makes sense that males would evolve to be able to invest to a larger extent in the well-being of their direct descendants.

Although the possibility for lactation in male mammals is tantalizing, not enough is yet known about how it functions. A system for which we have an abundance of data is that of the Syngnathid fishes, who are best known for being sex-role reversed. A certain level of sex-role reversal is observed in insects, amphibians and birds, but nowhere is it more clearly developed than in the seahorses and pipefishes. Here, males get pregnant and give birth to live young. Males of all species within the Syngnathidae receive unfertilized (yolk-rich) eggs from females in what is essentially their act of copulation. The eggs are then fertilized and stored by males in a wide range of structures from simple attachments on his ventral epithelium (belly), to partial cover by extended epidermis, to highly complex brood pouches.

Seahorses in the genus *Hippocampus* possess among the most complex brood pouches, and we can say beyond a shadow of a doubt that these males actually get pregnant. Far from being a mere storage area for eggs, brood pouches (and the males that possess them) have important functions in aeration, protection, and osmoregulation of eggs. The male is also able to physiologically transfer nutrients to developing offspring, indicating that his investment in a brood is substantial. This makes sense from an evolutionary perspective for the key reason that females transfer their eggs to males *unfertilized*. Having 100 per cent certainty of paternity appears to make all the difference in promoting the evolution of sophisticated paternal care in the Syngnathidae. In mammals, mothers provide immune factors to developing offspring through the placenta. This is called trans-generational immune priming, and it's critical for embryos to be able to fight off microorganisms and other pathogens once they have been born. The most sophisticated Syngnathid brooding chambers have additional tissues to provision developing embryos with nutrition and with complex immune factors, akin to the function of a mammalian placenta.

On one hand, sex-role reversal should lead to fairly predictable ecological scenarios. Since it's the males that are undertaking a huge investment in brooding the offspring, males should generally be the choosy sex and females should be competing for their affections. In several instances this is exactly what is observed: females of many species have complex and remarkable structures that are clearly the result of sexual selection. Adornments can take the form of skin folds or evaginations, larger body size, coloration, patterns on the body and elaborate courtship displays. As well as taking an active role in courting males, female Hippocampids take an active role in deterring other females, indicating that competition for males can be intense. However, there's an extremely important part of the puzzle that we cannot forget when discussing the relative investments of male and female Syngnathids: females are still contributing yolk-rich eggs. So in addition to having to compete for males' affections through either deterring other females or having a well-developed ornament of some kind, the ladies *still* have to produce the expensive gamete. This could be why we do not observe sexual structures with the complexity and extravagance of those observed in the avian world (think peacock). After all, a female has only got so much energy to work with. It is thought that females of the highest quality are those that can both display to males and escape from predation, without negative impacts on their fecundity. Females of lower quality are simply unable to juggle this extremely large set of biological demands.

We know that female birds can influence egg volume or clutch success based on the quality of the father, and male pipefish do something similar. In laboratory experiments designed to assess a male's investment in his brood after mating with high- versus low-quality females, male gulf pipefish (*Syngnathus scovelli*) demonstrated a clear ability to influence aspects of brood health. They are able to adaptively affect their investment in broods, depending on the value of the mother who contributed the eggs. Males paired with high-quality females allow for a greater number of eggs to be transferred to them, and the offspring realize a greater probability of survival to parturition (birth). Conversely, males paired with low-quality females experience higher rates of abortion (through re-absorption of tissues by the father). Since males are known to actively transfer both nutrients and immune factors to developing embryos, we can safely assume that they can also control the extent to which they withhold such compounds when faced with an unattractive brood. Males of sex-role-reversed species are more likely to save their energies for investing in high-quality broods, in the same way that females of traditional species do. So there you have it. Although the bulk of the animal kingdom is biased towards a larger role for females in the development and care of offspring, we can take heart in knowing that male Syngnathids are stepping up to the plate when it comes to providing their fair share of 'maternal' care.

Extreme lactation

Organisms that physiologically produce and provide milk to their newborn offspring are defined as mammals. It's thought that the provision of milk to young facilitated the diversification of mammals in both physical form and habitat type because newborn offspring can be completely sustained through its production. As long as a lactating female has enough to eat, her offspring are completely taken care of through production of colostrum and milk. Lactation is the most energy-intense time period for a female mammal, although there is great diversity in the composition and supply of milk between different groups. In addition, mammalian mothers are quite different from avian (bird) mothers in that the latter can intricately regulate the amount and quality of food provided to each individual chick. Mammalian mothers cannot regulate milk flow to individual pups, which sets the scene for sibling rivalry (see 'Siblicide' section below).

We are quite familiar with the lactational habits of Eutherial, or placental mammals, where mothers (humans included) provide milk to offspring that is fairly constant in composition (apart from the initial colostrum). However, there are two other categories of mammal that also provide lactational support to their offspring. These are the Prototheria, or monotremes (echidnas and platypus) and the Metatherians, or marsupials (wallabies and kangaroos). Monotremes are the only egg-laying mammals, and although monotreme mothers possess mammary glands, they do not have teats. Instead, milk is supplied to offspring through leakage on an abdominal milk patch and additional nutrition is supplied in the eggs. It is hypothesized that milk secreted just *prior* to egg hatching functions to protect the eggs and hatchlings from microbial infection – which speaks to the fact that mammalian milk is not merely a food source. Aside from providing a complex suite of antimicrobials depending on the needs of the infant, milk is thought to play a role in the direct development of the offspring's digestive system.

I'm going to take some time to talk about the remarkable developmental process in the third group of mammals, the marsupials, because it differs so drastically from the process in both monotremes and placentals. It's a set of fascinating developmental processes that have evolved on a completely separate evolutionary branch from the other mammalian groups. To start off, the group names are somewhat misleading. Prototheria translates to 'first beasts', Eutheria to 'true beasts' and Metatheria to 'behind true beasts'. In fact I'm going to change my description from somewhat misleading to completely wrong, because these names imply that humans and other Eutherians are somehow better or more evolved. As with any question pertaining to the evolution of a particular species or group, the process does not imply a progression towards some kind of 'best organism'. On the contrary, animals of all phyla are sufficiently well adapted to their current habitats and ecosystems to be each at their own evolutionary success point. These points are always changing, and one kind of organism would not do better to evolve to be more like another unless there is an adequate biological or ecological reason for it. So to define the Metatherians (marsupials) as *behind* the Eutherians is ridiculous. The fact that the unusual features of marsupial reproduction are largely limited to kangaroos and wallabies living in Australia means that they are uniquely adapted to these specific environments.

Many placental mammals produce young that are termed precocial, or very well developed after a long period of gestation. Think about organisms like sea lions or horses, where young are immediately mobile and able to direct themselves towards the source of their mother's milk. However, there is tremendous variation in the Eutherians with respect to the level of newborn development, from precocial to atricial offspring. The latter are less well developed and require a greater level of parental care. Think about human newborns, which are most pathetic in pretty much every way imaginable.

Marsupials have newborns that are termed 'highly altricial'. They are extremely undeveloped, approximately comparable to a human foetus at around eight to ten weeks of gestation. When female wallabies and kangaroos give birth, they sit on the ground with their tails between their legs, facilitating the subsequent journey of their tiny offspring. After a very short gestation (in wallabies it's 26.5 days, in kangaroos it ranges from 31–36 days) the itty-bitty babies use their forelimbs to propel themselves into the world through the mother's urogenital opening. The neonates crawl, unassisted, upwards along the mother's abdomen to reach the entrance of her pouch. They then switch directions and crawl down into the pouch towards the teats, where they will latch on and not let go for the next several months. Just around the time of birth, mothers will lick the furry pathway between their pouch and their genital opening, perhaps to keep the passageway moist and clean as the neonate makes its journey. I have to take a short pause here and marvel at this amazing feat accomplished by neonates that are hardly more than fertilized embryos. Can you imagine a human newborn crawling its way to find a teat to latch on to? Much less a nine-week-old human foetus? Although marsupial neonates are not well developed, their sense of smell is early to mature, and it allows them to find the teats through olfactory cues.

Once safely affixed to a teat within the mother's pouch, neonate marsupials remain there for the next several months. Adult female marsupials have essentially traded the umbilical cord for a teat, and after first latching the milk-flow begins. Mothers initially supply small volumes of dilute milk to their newborns. Digestion remains underdeveloped at this early stage, so milk is comprised of oligosaccharides (sugars) with almost no protein or lipid content. The lungs of the neonates are only partially developed, with gas exchange occurring through the rudimentary lungs and through gas exchange with the skin. These newborns are also ectothermic, so they obtain all of their body heat from their mother, and their immune systems are extremely under-developed. They are virtually unable to mount any kind of immune response until approximately one month after they enter the mother's pouch. This presents a conundrum: how do young with nearly no immunity manage to make their way to the pouch without being attacked by microbes or other pathogens? After all, the urogenital opening of females is hardly sanitary, and located mere millimetres from the anus. Over seventy-two types of bacteria have been identified from the urogenital opening of female tammar wallabies (*Macropus eugenii*), and a further fifty from the anus. There is also a great deal of bacterial diversity from within the pouch itself – which means that neonates must have some means by which to protect themselves from infection. Indeed, marsupial young are born with an outer epidermal layer called a periderm, which lasts for approximately one week and seems to serve an immunoprotective function as the neonates make their way to the pouch. There is a certain level of pre-natal transfer of immunoglobins to neonates from the mother, and a high level of immune compounds in the mother's milk. In addition, the pouch itself secretes antimicrobial compounds and both chemical and mechanical cleaning by the mother helps minimize the exposure of neonates to harmful bacteria and other pathogens.

The fact that the marsupial neonates are at such an early stage of development bears further discussion because their journey wouldn't be possible without the intensely specialized features of the mother's pouch. First off, it keeps mother and baby in direct contact with one another, which allows for a constant temper-ature to compensate for the lack of endothermic capability of the neonates. Mothers also control the level of humidity within the pouch, which facilitates efficient gas exchange through the newborn's skin. The level of gas exchange varies by species, in tammar wallabies it's a whopping 33 per cent of the neonate's oxygen supply at first, and it tapers to around 14 per cent at six days postpartum. As I mentioned earlier, antimicrobial com-pounds secreted within the pouch also help to confer some immune defence to newborns. On an overall level, the pouch itself is direct physical protection for the developing neonates from exposure to the elements, and to predators. It allows adults to exploit niches that are unavailable to organisms whose young are carried on their backs because it offers increased protection during jumping or gliding through difficult terrain. Not only are marsupial offspring well protected, they are inconspicuous – especially during their first phase of development where they are permanently latched to the teat and do not come out of the pouch.

In wallabies, this first stage of teat-latching lasts for approximately 100 days, after which the neonate sucks less often but remains confined to its cozy pocket. Starting at around 200 days after first crawling in, the newborns begin to intermittently leave for short periods of time. The milk composition of the female drastically changes from 200–320 days, with a much lower concentration of oligosaccharides and much higher concentrations of proteins and fats. Lactation in marsupials is a lengthy process, and one that is multifaceted: females tailor-make milk depending on the specific developmental stage of the young. Interestingly, the mammary glands of tammar wallabies function independently of each other, such that females can simultaneously feed a newborn pouch offspring milk of a completely different nutritional composition than that consumed by an older, out-of-the-pouch juvenile.

This highly unusual developmental strategy of marsupials may have evolved to provide parents with the option to terminate a pregnancy if environmental conditions become unfavourable. In the Australian outback, where habitats are unpredictable and extreme (i.e. irregular drought), selection for a developmental strategy where 'not much is lost' could be strong. Marsupials have a rapid birth, and a very low foetal cost in terms of time and energy. In addition, mothers retain the ability to resorb reproductive material following termination of a pregnancy, so the costs of losing an infant are extremely small. Moreover if a pregnancy is terminated or a neonate is lost, marsupials can almost always make up for the lost infant by producing another one immediately (see 'Arrested Development').

There are so many remarkable aspects of the developmental process of marsupials. Far from being 'pre-Eutherian', marsupial reproduction and development illustrate the power of evolution and niche expansion and unpredictable habitats. It's likely that such strategies haven't evolved elsewhere because there simply has not been a need for them. Having gestated and birthed four (utterly altricial) humans, it continues to astound me that neonatal marsupials exhibit such remarkable and complex behaviour when they are so tiny.

Set yourself up
for success

Human mothers would generally agree that we try to do the best job that we can for all our children. Care and resources are usually provided in equal amounts to all children within a given family, and picking a favourite offspring is something that is frowned upon (although at times I'm sure that most children would concur that some relative favouritism takes place). This is a pretty common theme in human families for a few reasons: first, our moral and ethical standards are such that providing equal care to all children is more important than provisioning them based on their potential for future contributions to our collective fitness (read: just because Suzie is smarter than Bobby and is more likely to secure a high-paying job as an adult, Bobby still deserves to eat). Second, there is a high level of paternal certainty in monogamous couples with children, meaning that the genetic makeup of all offspring is likely to be similar. As we know from previous sections in this book (see 'Monogamy. Really?'), the same cannot be said for couples in the animal kingdom, where paternity is likely to be variable both within and between mating seasons.

So what am I getting at here? To be perfectly blunt: sometimes, animal mothers pick favourites. They do so based on several factors, but a major one is the quality of the father that created them. One of the best ways to examine the variability in maternal investment is through studies on bird eggs. Eggs are these lovely little packages of maternal gifts. In addition to the embryo itself, the yolk and albumen have the potential to be variable in terms of their volume and also in terms of the myriad of other provisioning compounds that mothers provide. Things like hormones, antibodies, antimicrobial substances and antioxidants are all important parts of the egg that are under a mother's direct control. Once an egg has been laid, it's a done deal: no further provisioning can take place until the chick hatches out into the real world. The amount and quality of resources that a mother provides to an egg therefore has a massive influence on the behaviour, growth and survival of the progeny. This provides biologists with a detailed glimpse into how mothers can vary the love provided to different offspring under different conditions.

As I mentioned earlier, one of the factors with the greatest variability when it comes to offspring value is the quality of the father. One of my favourite examples to illustrate this point is that of the blue-footed booby (*Sula grantii*). The gorgeous blue-green feet possessed by the males are sexually selected – girls prefer boys with vibrant coloured feet to those with dull ones. Foot colour is a dynamic, condition-dependent trait, meaning that the current health and nutrition status of a male is easy for a female to assess. Foot colour can change in less than forty-eight hours if a male is deprived of food, so this characteristic is essentially a way of letting females know that a male is healthy and ready to be the father of her babies. In addition to this clear visual indicator of a male's fitness, the reproductive biology of blue-footed boobies provides an opportunity for biologists to manipulate it. You see, once a male and a female have copulated, the female will lay two eggs. The second egg will come along approximately four days after the first. Biologists interested in the ability of mothers to manipulate the level of resources to offspring have gone so far as to paint the bright blue feet of fathers to a dull grey colour between the laying of the first and second eggs. When they do this, the second egg comes along with an overall lower size and volume than the first, indicating that the mother is not going to invest too heavily in offspring that come from a substandard father. Egg volume and size are directly related to chick survival.

Termed the *differential allocation* hypothesis, the phenomenon of mothers' varying egg components in response to fathers' characteristics is well documented for many birds. For European starlings (*Sturnus vulgaris*) it's not about a male's direct phenotype per se, it's about an extension of his phenotype: his nest. Male starlings build their nests using twigs, old leaves and grass; they also utilize a combination of fresh green plants and herbs with aromatic properties. Females are more attracted to males that incorporate the fresh greenery into their nests, and also to males that are carrying plants and advertising the aromaticity (scent). There are direct benefits to raising offspring in nests with volatile plant compounds. Specific herbs positively influence several aspects of nestling growth and blood physiology, and these same greens inhibit bacterial growth and mite infestation within the nest. Indeed, female starlings show an increased allocation of testosterone to eggs laid in nests with a high level of green plant material. Yolk testosterone is thought to be beneficial in enhancing the growth rate and competitive ability of chicks.

Females of different bird species vary the levels of testosterone and other androgens based on many aspects of a male's attractiveness including phenotype and social dominance. The effects of androgen allocation to eggs are complex and can be difficult to interpret. As I mentioned above, in the short term, boosts of androgens are beneficial to developing embryos because they realize an increased growth rate and competitive ability. Long-term effects of androgen deposition may be negative, and sex specific. Male collared flycatcher (*Ficedula albicollis*) chicks exposed to experimentally increased egg androgens experience a decreased level of biological fitness. A lower survival rate of androgen-spiked chicks is facilitated through increased risk-taking behaviour and aggressiveness in addition to negative effects on general physiology. Androgens have immunosuppressive effects, leading to a lowered ability of supplemented chicks to stave off infections or parasites.

If the long-term effects of androgens are so negative, why do female birds of so many species deposit androgens in their eggs as a response to attractive fathers? This is a question that remains largely unanswered, although the need for short-term boosts in competitive ability may overshadow the long-term effects. After all, the most immediate requirement of any chick is to simply survive to adulthood. As evidence continues to accumulate about differential allocation and hormones, a better answer to this query will undoubtedly come available.

The contribution of other compounds, like immune-enhancing pigments or antimicrobials, is less controversial and paints a clearer picture of a mother's intent towards specific embryos. Male and female Eurasian reed warblers (*Acrocephalus scirpaceus*) are monomorphic: they are identical to each other. So how's a male going to show his prowess to a prospecting female? In this case it's all about the song. Female egg investment here varies according to the song complexity of males. When mated to males with highly complicated songs, females provision eggs with a higher level of egg-white lysosome – an antimicrobial compound that is important for the immune capabilities of the chick. Similarly, female canaries (*Serinus canaria*) lay eggs with heavier yolks and overall larger volume in response to experimentally exaggerated, highly attractive songs. Birdsong is an important sexually selected trait that provides females with information about a male's age, status, early developmental history and learning ability, and all of these characteristics are critical to the level of investment she should provide to the chicks of a given father.

In addition to male quality – which is certainly one of the most important determinants of differential allocation – females also provision their eggs according to other aspects of their immediate ecological situation. For example, predation risk is a factor that can have massive implications on how a mother equips her eggs. When female great tits (*Parus major*) are exposed to a high level of predation risk during egg-laying, their chicks tend to be smaller and lighter at hatching. Furthermore, their wings exhibit a faster rate of growth, perhaps indicating an improved ability of these chicks to escape from predators. The mechanism for such chick provisioning may lie with the level of testosterone allocated to the eggs, although it may not be as simple as providing chicks with a boost for growth and development. Instead, testosterone may act as a moderator of various developmental processes. Because the testosterone-provisioned chicks exhibit both a lower hatching weight *and* a faster wing-growth rate, it could be that this hormone can act to modify the specific growth pathways that are most important to the ecological situation into which the chick is hatched. The complexity of egg provisioning, especially when it comes to hormone investment, is mind-blowing, although we cannot rule out the fact that since mothers are exposed to predation and are likely to have increased corticosteroid levels within their own bodies, the transfer of steroids to eggs could also be a byproduct.

Breeding density is another factor that females take into account when provisioning their eggs. In a high-density situation, chicks will naturally be faced with higher levels of competition for resources and the potential for aggression from conspecifics or nest mates. Sociable weavers (*Phileairus socius*) provide an intriguing study system to examine the effects of differential allocation and density because these birds live in communal nests that vary greatly with respect to size. They can be huge, with over a hundred individuals all nesting together, or they can be small, with only a few. Females tend to provision eggs with increased levels of yolk androgens in crowded situations, which again speaks to the immediate benefit of increased aggression and competitive ability in chicks that have them. The bottom line with this, and all examples of differential allocation, is that mothers have a great deal of control over the kinds of provisions with which they provide their eggs. Far from being only about sharing gametes (as is the case with males), female birds are responsible for creating a care-package for their developing babies that has to reflect both their immediate and future ecological needs, as well as her own. This is an exceedingly complicated process that biologists have only scratched the surface of understanding. The fact that there are conflicting reports for almost all cases of egg provisioning (i.e. some females increase androgen levels, others decrease them) shows that there is much to this puzzle that we have yet to comprehend.

Take my kids. No really, TAKE THEM!

It may seem counterintuitive for an animal to put energy into creating offspring only to abandon them to the care of a stranger. After all, how can you make sure that your little ones will have the proper care that they require to grow up healthy and strong? However it's not unheard of for organisms, particularly birds, to give up their eggs to pseudo parents. It's called brood parasitism, and it describes a situation in which birds of one species lay their eggs in the nests of another. The host 'parents' go on to raise chicks that are not only unrelated – they are not even the right species. So, how does this make sense for either the parasite (the one that deposits the eggs) or the host? It turns out that this *can* be an extremely effective way of maximizing biological fitness, at least for the parasite. Think about it – once you've laid the eggs your parental investment is finished. For birds this is saying quite a bit, since most chicks require a good deal of parental care in the form of food provisioning. Indeed, in nests of parasitized species both mum and dad work extremely hard to bring back enough food for their whole brood (even though their brood contains some offspring that are, somehow, different).

Once their eggs are gone, parasite parents can then go ahead and put more effort into making more babies. This means that there is a distinct advantage for parasites, and a distinct disadvantage to the hosts, setting the scene for an evolutionary arms race between them. Hosts should evolve mechanisms to rid their nests of parasitic eggs, and parasites should then evolve ways to evade those mechanisms. This is pretty much exactly what we observe in the bird world, though there is tremendous diversity with respect to the specific strategies of hosts and parasites, and also with respect to the number of species involved. Some are generalist parasites, like the common cuckoo (*Cuculus canorus*), a parasite that lays eggs in the nests of more than a hundred different species. Others are completely specialized to lay eggs in the nests of only one species, such as in African indigo birds (family Viduidae) where host specificity is extremely high and parasites have evolved to have chicks that look and sound similar to those of their hosts. In addition to this diversity in host specificity, there is also diversity in the frequency of nest parasitism, the number of eggs provided to different nests, and the strategies of parasitic chicks once they hatch.

Parasite chicks can be either evicting or non-evicting, meaning that they do or do not toss out all of the existing eggs or nestlings that are in the nest with them. Obligate evicting parasites effectively destroy the reproductive success of their host parents by killing all their offspring (not to mention that the parasitic young go on to utilize the resources brought home by the parents). Conversely, the bulk of brood parasites are non-evicting, meaning that they don't turf out the offspring of the host, though they often out-compete them for the food resources because they are usually larger and more aggressive.

Either way, it seems that host species get the raw end of the deal when it comes to brood parasitism. Many have evolved strategies to identify and rid their nests of the unwelcome additions, both before and after they arrive. For example, chalk-browed mocking-birds (*Mimus saturninus*) have adopted a mobbing strategy as a front-line defence against their shiny-cowbird (*Molothrus bonariensis*) parasites. Mobbing can include swooping close to the parasite, attempts to knock it from the nest, chasing and biting or striking it. More than one individual may join the fight against the intruding parasite, although unfortunately it doesn't generally prevent the parasite from laying an egg. How come? Well, shiny cowbirds are extremely efficient parasites, having evolved strategies of agile flight, opportunistic timing and ultra-fast egg-laying. They wait for the right moment to swoop into a nest of their mockingbird hosts, usually when the hosts are away or are engaged in mobbing with other cowbirds. Sometimes they use a tailing strategy, where two cowbirds fly to a nest in quick succession, the first one bearing the brunt of the mobbing behaviour while the second one lays an egg. Egg-laying takes approximately seven seconds for shiny cow-birds (compared to the approximate twenty minutes observed for many bird species), and so they only need a very small window to be able to successfully deposit. However, all is not lost for the mockingbirds. When shiny cowbird females have more time on their hands they will use their beaks to poke holes in the eggs of the host species that are already in the nest. They do this during visits to the nest before laying their own eggs and also on laying day. When mockingbirds are engaged in active mobbing, the incidence of egg destruction is markedly lower. In other words, they cannot stop the parasitism from taking place, but they *can* stop damage to their own eggs by mobbing the parasites, which has benefits in terms of biological fitness. Mockingbird chicks are a similar size to cowbird chicks, and they are able to compete successfully for parental resources. So, if a mockingbird can keep its eggs safe from destruction prior to hatching, chances are that the chicks will survive. A similar scenario plays out between host yellow warblers (*Setophaga petechia*) and their parasite species, the brown-headed cowbird (*Molothrus ater*). When warblers take an attentive position at the top of their nest, they are less likely to experience egg removal by the cowbird, although this behaviour does not affect whether or not their nests get parasitized.

Other cowbird species have evolved complex systems of egg-mimicry to remain successful at brood parasitism. Female common cuckoo (*Cuculus canorus*) are quite selective about the nests that they parasitize; they make several observational trips to various nests prior to laying. What they are looking to do is to match their own eggs with the chromatic composition of the eggs already in the nest, so some host nests will be a better fit than others. Egg appearance is an important characteristic that could contribute to whether or not a host family accepts or rejects a brood parasite's egg, so selection has favoured the ability of cowbirds to chromatically mimic host eggs with a good deal of reliability. Conversely, hosts have evolved fine-tuned abilities of egg discrimination that allow them to identify foreign eggs and remove them from the nest. If there is more than one foreign egg present, parents could choose to abandon the nest altogether and start another one in a different location.

For host great-reed warblers (*Acrocephalus arundinaceus*), the timing of foreign egg deposition is an important factor in determining whether or not to reject it. Warblers reject common cuckoo eggs approximately 30 per cent of the time, and the probability of rejection decreases as the host's eggs near hatching. It's likely that once the eggs of the host species are late enough in development they will hatch before or at the same time as the brood parasite, and chick survival is no longer compromised.

In this scenario it may take too much energy for a host parent to fend off the continuous stream of brood parasite eggs than to accept a few of them.

So many aspects of these relationships have the potential to be fine-tuned by both parasites and hosts. Acoustic structure of a chick's begging call is another characteristic under a good deal of evolutionary pressure. The calls of shiny cowbird (*Molothrus bonariensis*) chicks contain repeated syllables rather than one simple chirp, as is the case for its host, the fairy wren (*Malurus cyaneus*). These multi-syllabic calls are termed 'tremulous' calls, they function akin to a repeated form of the host chick call and they are quite distinctive from the calls of wren chicks. Interestingly, host parents respond to the call of the shiny cowbird by ramping up their provisioning activity. They do this because the tremulous calls exploit a common provisioning rule in birds: repeated signalling to parents over a shorter time frame results in a greater amount of food. In other words, the squeaky wheel gets the grease.

Although the scenarios depicted so far paint a dismal picture for the host species, there may be times when hosting a brood parasite could be advantageous. For example, it's speculated that host carrion crows (*Corvus corone*) can benefit from the presence of great spotted cuckoo (*Clamator glandarius*) chicks, which would make this a more mutualistic scenario than a parasitic one. The great spotted cuckoo is a non-evicting parasite that is fairly specialized to corvid (a large bird group including crows, ravens and jays) species. If the host eggs make it to hatching without being lost or damaged, crow chicks can successfully compete with cuckoos so their survival probability is high. Interestingly, crow nests that are parasitized have a lower failure rate than non-parasitized nests when predation pressure is high. This is probably due to the fact that the cuckoo chicks secrete copious amounts of a noxious fluid when grabbed by predators (usually large raptors or feral cats). This cloacal secretion contains a suite of acids, phenols and sulphur – and it's very effective at deterring predators. So it follows that a modicum of protection is offered to crow chicks when the parasitic stink-bombs are around. However, I should stress that the mutualistic trend in this relationship only stands when predation pressure is high. At low levels of predation the greatest nest success lies with crows that are not parasitized.

A further addition to the world of brood parasitism is that it can sometimes occur within a species. Termed conspecific brood parasitism, this seems like a more common-sense option because closely related mothers can still realize an increase in biological fitness despite providing care to offspring that are not their own. There is some evidence that close relatives more frequently parasitize each other, although there is also evidence to the contrary. In goldeneye ducks (*Bucephala clangula*), multiple same-species parasites deposit eggs into a single nest. The host female is therefore forced to care for offspring from several other mothers. Female goldeneyes exhibit high-site fidelity, meaning that within a given area the occurrence of female:female relatedness is high because they don't stray too far from the place where they themselves hatched. While it has yet to be demonstrated whether parasitic females lay eggs in the nests of closely related females, it *has* been demonstrated that closely related parasitic females will lay eggs in the same host nest. Sisters are more likely to parasitize in the same place, keeping their kin together. They are also more likely to select host nests that are located in areas that are less exposed and prone to predators. To put it another way: if you're going to abandon your baby, at least abandon it with a cousin, and do so in a safe neighbourhood.

So far this discussion has focused entirely on the existence of brood parasitism in birds, which makes sense seeing as this is where its occurrence is most widespread. However, brood parasitism does occur in other organisms as well, such as fish. The mochokid catfish (*Synodontis multipunctatus*) is a brood parasite on several mouth-brooding cichlid fishes in Lake Tanganyika. Mouth-brooding females oviposit on the substrate, and then quickly scoop the eggs up into their mouths either before or after fertilization by a male (depending on the species). Eggs are incubated in the female's buccal cavity (mouth), and they remain in this safe environment until their yolks have been fully absorbed. Once out of their yolk-based food, tiny hatchlings leave the mother's mouth to forage, but they return for refuge when not foraging until they become large enough to live independently. It remains unclear how eggs of brood parasites make it into a female's mouth, it's possible that the parasite females sneak a few of their own eggs on to the pile while host females oviposit. However, one thing is certain: eggs of the parasite are incubated alongside those of the cichlid. Unfortunately, the parasite eggs hatch prior to those of the host, which results in the parasite fry feeding on the eggs and small cichlid fry while still in the mother's mouth. There is much more to be learned from these systems. What kind of selective pressures would lend themselves to the ultimate scenario of a mouth-brooding female losing an entire clutch while providing such substantial benefits to a foreign juvenile? It's a question that has yet to be answered.

Overall, it's thought that the success of generalist brood parasites is on the rise due to their high reproductive success and adaptability. If you're a parasite that has evolved the luxury of depositing your eggs in any number of nests of several species, chances are you will be reproductively successful regardless of long- and short-term environmental stochasticity. Generalist brood parasites tend to have fairly stable populations seeing as their bet-hedging strategies allow them to spread out their risk of nest failure. This could mean that in the era of global climate change and associated environmental modification these organisms, unlike many others that exhibit a high degree of parental care, will thrive.

You should not
kill a child

It's the ultimate biological slap in the face. Once you've finally managed to meet a partner, have sex and create offspring with said partner . . . someone comes along and kills the fruits of your labour. It's an awful prospect, and those of us that are parents can hardly imagine such horrors. Unfortunately, in the animal kingdom infanticide is another of those things that happens a lot more than we'd like to believe. The killing of offspring occurs in diverse animals. In fact, it has been observed in over a hundred mammalian species (including some of our closest primate relatives and our own species). In several cases infanticide is common enough to be one of the major contributors to juvenile mortality overall.

The bulk of infanticide perpetrated by adult mammals is thought to relate directly to a dominant male's biological fitness. The 'sexual selection' hypothesis for infanticide explains the fairly common scenario of a male killing offspring sired by another father. This kind of behaviour is almost always exhibited in polygynous species were one male is more dominant than most others, like in primates, lions, horses and bears. Dominant males kill offspring sired by other males as a way of increasing their own biological fitness, as I have described extensively in previous sections of this book. The unfortunate reality for females of diverse species is that at least some of their offspring will be killed by an adult male, and that same male will then aim to create new babies with them. In most cases there is absolutely nothing they can do but comply.

In some species where offspring are subjected to sexually selected infanticide by males, females *have* developed counter-strategies such as being aggressive towards males that come close to them and their offspring, keeping offspring in hidden burrows (rodents) or having them cling to their own body (primates). Females of several mammalian species also make the choice to mate with the most aggressive, dominant male in the first place. They are then automatically protected from his infanticidal behaviour, and they are most likely to be protected against attacks by other (less dominant) males. In other species females confuse paternity by copulating with several males during their fertile phase.

Males are not the only sex to engage in infanticide. The ugly truth is that in many cases it is a female causing the death of another female's babies. One of the most common reasons for infanticide between females is explained by the 'resource competition hypothesis', which suggests that the murderer (and her descendants) will have improved access to limited resources like food, shelter or territories. In cooperatively breeding mammals like wolves (*Canis lupus*), mongooses (family Herpestidae) and meerkats (*Suricata suricatta*), infanticide between females is common, and is best explained in the context of resource competition. For example, many female rodents find themselves competing with each other for access to suitable territories. The easiest way for a female to 'win' a competition for a territory is to kill all the offspring of her competitor, who will then leave the area. Female reed warblers (birds) are also in competition with each other, but in this instance it's for an increased investment from the father of the offspring. Secondary females are subordinate 'squatters' on a primary female's territory. The same male fertilizes both primary and secondary female warblers, but the secondary will not receive as much help in raising her offspring as the primary. Unless of course the secondary female happens to destroy all the eggs in the primary female's nest, which serves to direct the father's attention back to her.

In mammalian social groups, there are several females present; however, usually only one of them is socially dominant. This single female realizes substantially higher reproductive success than the others, who take on a subordinate social role and generally help rear her offspring. In meerkats (*Suricata suricatta*), subordinate females have a lower reproductive success than dominants both because of the fact that their access to sexually mature males is limited and also that the dominant female will immediately kill their pups once they are born.

Interestingly, the subordinate meerkat females also do their fair share of baby-killing, becoming especially murderous when they are pregnant. Any litters that are born (by other females) during their pregnancies may be killed and consumed – regardless of whether they are litters of another subordinate female or of the dominant female. It seems that pregnancy brings out the worst in female meerkats. Dominant females have evolved another strategy to deal with the infanticidal behaviour of the subordinates: for the last three weeks of their pregnancies they banish all subordinate females from the group. Those that are not pregnant at the time are unlikely to conceive once they are driven away from the group, and those that are pregnant at the time are more likely to suffer a miscarriage due to stress. Once the subordinate females have been driven away, the dominant female can give birth and enjoy a lower rate of infanticide from the subordinates.

Although the sexual selection hypothesis and the resource competition hypothesis explain the majority of infanticide in mammals and birds, there are several other hypotheses to explain alternative scenarios. The 'adoption avoidance hypothesis' suggests that infanticide is a pre-emptive measure that's undertaken by parents to avoid spending valuable resources on offspring that are not biologically related. This kind of situation is more likely to be observed during periods of resource shortage where parents are barely able to feed themselves and their own offspring, making them more vigilant about accidentally sharing with others. For example, sea birds often nest in large roosting colonies where infants are left for periods of time while the parents head out to sea to gather resources. When times are tough and both mother and father are required to spend long periods searching for very little food, they want to ensure that it's their own biological offspring who are receiving the goods. This includes killing the offspring in neighbouring nests. Similarly, in breeding colonies of seals and sea lions, young pups whose own mothers are either foraging or (during tough times) have died, will attempt to suckle milk from neighbouring mothers. They attempt this while the mothers are asleep because such behaviour is not treated kindly if she becomes aware of it. Adult females will often kill pups that are trying to 'steal' milk in this way.

The 'predation' hypothesis suggests that adults gain nutritional value from consuming the young of others. While it may not be the primary reason for infanticide, the consumption of killed offspring is observed in most mammalian species that undertake it (a fringe benefit, if you will). In rodents, an increased incidence of infanticide is observed for males during periods of food deprivation, and for females during periods of lactation (which confers high energetic demands). Lastly, the 'social pathology hypothesis' suggests that in some cases infanticide is simply a mistake. It may be more likely in habitats that have been recently disturbed or in especially crowded or stressful situations. In other words: sometimes, animals kill their own or each other's babies, and we don't know why.

I've kept the discussion focused on vertebrates so far, but the occurrence of infanticide is widespread in the invertebrate world too. It doesn't necessarily fall into the same categorical hypotheses as we observe for vertebrates, and this is because there are striking differences between vertebrates and invertebrates with respect to individual recognition. As biologists, we can say without a doubt that vertebrates have the ability to discern their own offspring from those of conspecifics. With invertebrates we can infer this at times, but it's much more difficult to prove. Due to their larger size it's possible to classify specific individuals and behaviours in vertebrates, whereas human observers struggle to do the same when it comes to invertebrates. So there are likely to be distinct differences in what we are able to observe about invertebrate infanticide, and also the degree of discrimination used by invertebrates when it comes to committing the act itself.

An example with burying beetles (*Nicrophorus vespilloides*) illustrates this point nicely. Among invertebrates, burying beetles have an exceptional level of parental care. Male and female parents prepare and bury small vertebrate carcasses, which will serve as the sole food source for both parents and offspring during the phase of larval development. The discovery of the carcass is the first step in the process towards becoming beetle parents, triggering oviposition in females. In addition to preparing the corpse as provision for their offspring, parents also regurgitate food for their hatched infants for the first two instar (developmental) stages. As I mentioned earlier, this is a high degree of parental care, and so parents have evolved mechanisms to ensure that they are providing said care at exactly the right time. These parents rely on something called 'temporal kin recognition', which means that they recognize their own offspring based on cues related to time (rather than visual or olfactory signals). Prior to being in 'parental care' mode, both mum and dad beetle are in 'infanticide' mode, whereby they are clearing their carcass of other larvae by eating them.

Starting from the occurrence of oviposition, the timing of embryo incubation is dependent of photic, or light cues. Day-length inputs provide the specific signals to the beetles about when they need to *stop* eating larvae and *start* taking care of it. This is generally fifty-seven hours after oviposition, although mum and dad cease to consume larvae for a period of eight to twelve hours before their own eggs hatch. In this way, beetle parents are successful at parenting, and successful at infanticide (in terms of biological fitness) without actually being able to physically recognize their own offspring from anyone else's. Invertebrates have evolved a myriad of techniques like this to ensure high levels of biological fitness.

I'll end this section with an example that extends what we mean when we talk about the killing of other's offspring. Infanticide need not always be intraspecific. Although you may at first want to classify infanticide between species as predation (and you'd be right to do so), there are some examples from the invertebrate world that require further explanation. What happens when adults of one species kill the infants of another, and they are not gaining any nutritional benefit by doing so? Omnivorous thrips (storm flies, order Thysanoptera) often cohabit on the same plants as their biggest predators, several species of mites in the family Phytoseiidae. The mites are predatory on the adult stages of the thrips; however, the juvenile stages of the mites are not predatory, and they are vulnerable to infanticide by the adult thrips. The killing of the mite larvae has the immediate effect of reducing the predation risk to the adult thrips because adult mites are four times less likely to oviposit where their eggs were killed. That was a mouthful, so let me say it another way: when their babies all get killed in a certain area, the predatory mites are far less inclined to lay more eggs in that specific spot. This is excellent news for the thrips, which realize a much greater survival rate when their predators are deterred.

This same 'infanticide' occurs regardless of whether the thrips and mites are living on either highly nutritious or non-nutritious plants. This indicates that the adult thrips are not committing infanticide on the infant mites because they require additional nutrition, it's essentially a counter-attack to the predation that these adults would face if these infants were to grow up. This kind of reverse predator/prey relationship is intensely interesting because it takes into account the fact that even the most voracious predators must first pass through a stage in their lives where they are as vulnerable as an adult prey species. The effects of the developmental process on the ecology of a species should not be underestimated.

Child abuse

Humans have invented all manner of strategies for keeping our offspring safe while we are away. In an optimal world parents would never have to leave their children, but the cold hard truth in our species and many others is that in order to provide enough resources, foraging away from home is a necessity. In organisms that breed in large groups the costs of both parents being away from the offspring can be mitigated by alloparents, non-breeding kin that lend a hand in terms of protection and feeding. Alloparenting is more likely to occur peacefully when resources are abundant and the actual parents can pay back the alloparent and minimize time spent away from their babies. However, when times are tough, the situation tends to go wrong on a few levels. Parents are likely to spend a greater amount of time foraging for a smaller amount of food, and alloparents can turn infanticidal.

This is a common scenario among sea birds, where breeding and roosting areas are often a healthy distance away from oceanic environments where parents forage for food. Nazca boobies (*Sula granti*) are ground-nesting colonial seabirds that exhibit highly unique pseudoparenting behaviour. In any given year there are adults within the colony who do not breed. These are either males who have not found a mating partner (their sex ratio is generally skewed) or females whose nests have failed. While parents are away foraging, these nonbreeding adults (NAVs) pay 'visits' to unattended chicks, hanging out with them for periods of between one minute and an hour. Newly hatched chicks and almost-fledging chicks are *not* visited by NAVs because parents almost always guard new hatchlings, and chicks near fledging can defend themselves. It's the chicks that are in the middle that receive the visits, they are too young to defend themselves but too old for their parents to remain by their side. During NAV visits chicks will assume a submissive posture, with their heads inclined forward and the dorsal surface of their beaks on the ground. It's thought that this subordinate posture may help to protect their eyes, but overall it doesn't seem to accomplish much.

NAV visits can take a few different forms. Some are affiliative (friendly) and peaceful: NAVs will stand beside or preen young chicks, or in some cases even provide them with gifts of pebbles or feathers. Unfortunately, the bulk of NAV visits involve aggression by the visitor, who will scratch or bite chicks – leaving them with bloody wounds on their necks. The scratches aren't fatal, but they attract the attention of blood-sucking ectoparasitic birds like finches or mockingbirds. These ectoparasites will feed from the cuts and deepen them substantially, which often leads to the death of the chicks.

The last form of the NAV visits is the most perplexing: adults (usually males) will engage in sexual copulations with chicks. Biologists have yet to come to any kind of consensus as to why this occurs. The bottom line is that all three kinds of NAV behaviour in Nazca boobies are extremely common. It's observed across different colonies in different areas, and therefore cannot be dismissed as some form of behavioural mistake. Up to 24.6 per cent of chicks die as a result of the direct or indirect effects of NAV behaviour.

It's possible that non-breeding adults use the chick-visits as an adaptive strategy to acquire a nest or a mate if the chick's injuries result in the ultimate divorce of the parents or abandonment of the nest, although this seems like a roundabout way to achieve that. Other reasons for NAV behaviour include the possibility of eliminating future competition from the mating pool (if the NAV behaviour is directed towards a chick of the same sex), or reducing competition for resources. However, none of these hypotheses explains the fact that NAV visits can also be friendly (affiliative) or sexual. As yet there isn't a well-supported explanation for the existence of NAV visits in Nazca boobies.

During maltreatment nestlings experience a five-fold increase in corticosterone, and levels remain high for several hours post-trauma. The physiological effects experienced by chicks as a result of repeated abuse appear to cause irreversible changes in the hypothalamic-pituitary-adrenal axis of the brain, causing long-term neuroendocrine changes that result in the establishment of future NAV behaviour in the chicks. Basically, in something aptly termed the 'cycle of violence', chicks that are heavily abused are themselves hard-wired to become abusers when they reach adulthood. There is also a positive correlation between adult abusive behaviour and siblicidal experience. Chicks that manage to survive a siblicide attempt experience a sudden and intense up-regulation of testosterone. As adults, they are more likely to engage in NAV behaviour. On an overall level, what this means is that chicks that are abused tend to become abusers. Despite the fact that the NAV phenomenon is localized to Nazca boobies, the clear demonstration of abuse and its aftermath warrant further study and description. It's entirely possible that this kind of behaviour is more common in the animal kingdom than we would like to believe.

Siblicide

Human families tend to exist in a state of relative cohesion. Although there is invariably minor bickering between siblings these squabbles tend to work themselves out in a peaceful manner. In the animal kingdom that is not necessarily the case. When competition for resources is high, there can be intense strife between siblings to secure a fair share. Such scenarios produce clear winners and clear losers – to the point of death. Siblicide is fairly common in several animal groups, although it occurs to a greater extent in many bird species.

In the early phase of their lives nestlings are entirely dependent on food deliveries by the parents, which sets the scene for competition between siblings when the mother or father arrives back at the nest. Chicks with more effective begging techniques are more likely to be noticed and fed, and for the most part effectiveness in begging results in acquiring more food and achieving a larger body size. Dominance hierarchies between bird siblings are an inevitable part of their existence, and they are usually dictated long before chicks even hatch. Social structure among bird nestlings is dependent on hatching order. Chicks that hatch earlier have a greater opportunity to grow (via parental-assisted food delivery) than do chicks that hatch later, and so the inevitable (and unfortunate) reality for later-hatching chicks is that they are destined to be picked on. Not just picked on either – they can be pecked on (literally) until they die, or they can be starved to death by bullying siblings who take all the incoming food from mum and dad. Parents rarely intervene in this kind of brood reduction; it is a normal and natural part of life in the bird world.

There are two scenarios surrounding siblicide in birds. Facultative siblicide describes situations where death occurs because of unusually low resources or some other stochastic environmental factor. When times are tough, it's all about survival of the fittest and the weakest siblings are picked off first. In species with facultative siblicide there isn't necessarily a large size difference between juveniles that hatch earlier versus later, and differences in competitive ability between hatchlings are fairly low. Furthermore, mothers retain the ability to modify the level of competitiveness of their hatchlings by selectively varying the level of androgen-based hormones like testosterone in the eggs. When times are good, mothers provision late-hatching offspring with an extra dose of testosterone to keep the competitive playing field relatively even between siblings. In this way, the later-hatching offspring can add extra reproductive value to a female when resources are abundant and times are good. However, when times are tough and resources are difficult to come by, such extra provisioning to late-hatching offspring does not happen, and they will most likely meet their untimely deaths by starvation or wounding by larger and more competitive siblings. It's a form of bet-hedging, or biological gambling, which occurs in many species and serves to increase biological fitness over the reproductive lifetime of an individual. Birds of diverse species from blue-footed boobies to sparrows and tits exhibit facultative siblicide and varying levels of maternal egg provisioning depending on the immediate environmental conditions.

The second kind of siblicidal scenario in birds is obligate, and is exhibited by much fewer species. Nazca boobies (*Sula granti*) and masked boobies (*Sula dactylatra*) display this kind of behaviour, where there are always two kinds of offspring: the core group, and the surplus group. The horribly unfortunate news for the surplus offspring is that they are just that: extra. They are meant to live only if some kind of stochastic accident or predation event kills one of their precious core-group siblings. In these species we generally see large size differences between early and late-hatching individuals, and mothers tend to provision early (not late) hatching eggs with extra testosterone.

As I mentioned earlier, the only way that a surplus offspring will survive is if some kind of random accident wipes out one or more members of the core group. Otherwise, the most dominant chick in the brood will kill the surplus offspring. It happens in every single brood. The major difference between the late-hatching offspring in facultative versus obligately siblicidal species is that in the former their role is to increase reproductive value, whereas in the latter their role is strictly as biological insurance.

Mammalian species with a high maternal investment (like hyenas, family Hyaenidae, and Galapagos fur seals, *Arctocephalus galapagoensis*) exhibit a similar scenario of facultative siblicide as that observed in the bird world. Young spotted hyenas are completely dependent on their mother's milk for the first twelve to eighteen months of their lives, although siblicide normally occurs during the first three months of life while a twin litter is still housed in an underground burrow. Hyena milk is very high in fat and protein, and it is physiologically expensive for mothers to produce. Cubs are generally born as singletons or twins, though triplets have also been observed. As I've mentioned in an earlier section of this book, juvenile hyenas are extremely violent and are born with strong neck muscles and sharp teeth. The potential for competition for mother's milk, and the clear fighting abilities and violent tendencies of newborn hyenas set the scene for siblicide. Crucial to the situation in this and any other facultatively siblicidal organism is the fluctuating availability of resources to the mother (which she can then turn into resources for the newborns).

The Serengeti plains can be variable with respect to the level of herbivore prey depending on the time of year. In a time of abundance a mother can easily provision one or two cubs (three is rare); however, when mothers must forage over hundreds of kilometres to find enough food, the situation drastically changes. Not only is she utilizing more of her own energy in travelling such large distances, she is not producing as much milk, and is away from the den for longer periods of time. During these difficult times there is more than ample opportunity for the most aggressive juvenile to kill his or her littermate. Interestingly, hyena mothers do not decrease their investment in offspring once an act of siblicide has taken place. Milk production and allocation remain the same, and so the surviving juvenile experiences a high level of reward for killing his/her nest mate.

As we learned in an earlier section, hyena societies are matriarchal, dominated by females. The social status of males plummets at sexual maturity; after which they are generally recognized as being the lowest-ranking members of society. These social roles are evident from a very early stage of life. The siblicidal offspring in twin litters are most often the females. In the rare instances where a brother is dominant over a subordinate sister, she is more often able to utilize effective counter-dominance tactics than is a subordinate brother against a dominant sister.

In a similar scenario, Galapagos fur seals (*Arctocephalus galapagoensis*) and sea lions (subfamily Otariinae) rear a single offspring at a time and nurse it exclusively for a period of two to three years. However, many females give birth to another offspring during this long phase of lactation, which creates a competitive situation between two very differently sized offspring. During phases of high resource abundance, mothers defend their newborns from intimidation or violent tactics of the older offspring, often thwarting their attempts to nurse. When abundant food is available the older offspring stand a greater chance at finding food of their own and surviving without the mother's milk, and she will favour this tactic in order to provide more milk to the newborn. Unfortunately, during El Nino years or other times when food resources are less abundant, the same inevitable end comes to the second-born offspring. Mothers allow for the intimidation and starvation of the newborn and continue to provide as much nutrition as possible to the older, more dominant pup. Since she's put a greater level of investment into the older pup, it makes sense for her to facilitate its survival over the newborn.

Siblicide is also rampant across the invertebrate world, occurring in diverse taxa in social species where workers or parents provision developing young. As with the bird and mammal examples, food provisioning by an outside party means that competition will develop between the juvenile organisms that require said food to survive. In many invertebrates the preferred mechanism of siblicide is cannibalism. Thus the benefits of siblicide in these animals are twofold: first, siblicide eliminates competition for parental-derived food, and second, the act of cannibalism itself provides a direct nutritional benefit. Now, if we think about the notion of biological fitness there is also a potential loss in terms of inclusive fitness if one individual consumes a genetically related sibling. We would therefore expect siblicidal cannibalistic species to evolve mechanisms to distinguish between kin and non-kin at communal nesting sites.

European earwigs (*Forficula auricularia*) are a sub-social insect species in which all life stages display aggression and have cannibalistic tendencies. Early instars are found in broods with a variable composition of paternal half siblings and full siblings (depending on how many males fertilized a female's eggs); however, later developmental stages exist with a greater mix of related and non-related nymphs due to immigration of nymphs from foreign broods. Larval earwigs adjust their nest-mate killing and cannibalistic tendencies according to what would be expected under the umbrella of kin selection: they kill and consume unrelated brood mates to a much greater extent than siblings. This means that cannibalistic individuals benefit from a direct nutritional meal, and perhaps also from a decreased amount of larval competition for supplementary food. Similar trends have been demonstrated in other arthropods such as ladybird beetles and some species of wolf spiders, as well as in several groups of amphibians.

Tadpoles of the spadefoot toad (family Scaphiopodidae) exhibit two distinctive strategies with respect to their nutritional needs. There are those that feed on detritus (decaying plant and animal matter) and later develop into omnivorous herbivores; these individuals have flattened mouthparts that are morphologically suited to herbivory. Conversely, there are tadpoles that feed on fairy shrimp and develop tendencies towards carnivory and cannibalism; these individuals have beak-shaped mouthparts and larger mouth-musculature that facilitates their predatory habits. Carnivorous tadpoles exhibit a solitary social scheme, as opposed to omnivorous individuals that tend to associate with conspecifics. Although both morphotypes *can* engage in cannibalism, the omnivorous individuals do so to a much lesser extent than the carnivores. It would follow that the kin-recognition abilities of this second morphotype should be much sharper than those of the first. Instead, both types of tadpole exhibit a refined ability to discern kin from non-kin. Why should this be? Well, the specific morphotype of any individual is environmentally determined. It can be reversed in a laboratory setting, and may even change through the duration of the tadpole stage. For these reasons it follows that the (genetically derived) ability to differentiate kin is present in all individuals, although the direct benefits to carnivores for avoiding kin are higher because omnivores don't tend to cannibalize in the first place.

The selective pressure to avoid cannibalizing close kin may not always be strong enough to warrant the evolution of strong means by which to distinguish them. South African praying mantis (*Miomantis caffra*) nymphs are another species in which larvae are reared in groups of related and unrelated conspecifics. They are aggressive hunters and strongly cannibalistic, and they do not distinguish siblings from non-siblings when it comes to their cannibalistic tendencies. In other words, they are as likely to snack on a sibling as an unrelated conspecific. This relates directly to the ecology of the mantis nymphs, who are generalist ambush predators and are frequently food limited. When food is scarce and starvation is a real possibility, cannibalism makes the most sense, regardless of whether you are related to your meal. In the ecological context of the South African praying mantis, selective pressure favours indiscriminate aggression over kin-selective siblicide.

Clutch piracy

The mention of the word pirate makes me think of heartless thieves, raping and pillaging their way around the world. This description isn't too far off the mark when it comes to a little-known phenomenon called clutch piracy. So far this kind of behaviour has only been observed in a few species of the amphibian world, although it is thought that it could be very common. The notion of clutch piracy is one of exploiting the reproductive efforts of others for personal gain. Some other parents have done all the groundwork, so a clutch pirate has relatively little to do in order to gain some reproductive success. For example, the Amazonian poison frog (*Dendrobates variabilis*) has two main stages in its development to adulthood. First, eggs are oviposited and fertilized (by mother and father respectively) and they are then placed by the male at the water/air interface of small pools or bromeliads. Once the eggs have developed into tadpoles, the father will scoop them up and transport them to a different pool where they will undergo development to the adult form. Both eggs and tadpoles are often found at the same locations, although they are rarely related. Tadpoles are very cannibalistic and aggressive towards each other, and those with a size advantage (older) are usually the cannibals. So fathers have a strategic decision to make when it comes to depositing both his fertilized embryos and his tadpoles.

Tadpoles in the pools can easily pick off embryos, especially if water levels rise and they become submerged; so fathers will avoid placing embryos at pools currently occupied by tadpoles. However, once a male has deposited his embryos, another male can take advantage of this free 'snack' by depositing his tadpoles into the same pool. In this way, adult males actively facilitate cannibalism of unrelated embryos by their own tadpoles. Because of the fact that males can directly manipulate the reproductive success of others by setting up their own offspring to cannibalize those of conspecifics, this certainly qualifies as an example of clutch piracy.

Another form of clutch piracy is found in common frogs (*Rana temporalis*) across the diverse landscapes of Spain and Romania. During normal reproduction males clasp a female's abdomen and the pair deposits a spherical clutch of eggs. This female-holding behaviour is called amplexus, and it allows the male to monopolize the reproductive efforts of the female while she's laying. Parental investment is over once the eggs have been deposited, and they are left to develop in their watery home. Often the fertilizing male will not manage to inseminate all of the female's eggs, leaving some unfertilized. This is where the clutch piracy comes in. Other males (perhaps those that have not had a chance to reproduce with a female) search for freshly laid clutches, and enter into a state of amplexus with *them* instead of with a female. Essentially, the male will be 'making love' to an egg mass with the intent of fertilizing some of the eggs that were left over. In this way, he can take advantage of the substantial reproductive efforts of the main male, while increasing his own biological fitness (and that of the female) by fertilizing eggs that would otherwise not develop. These frogs are explosive breeders, and the sex ratio is highly male biased, which means that not all males are going to mate with a female. Perhaps clutch piracy evolved as a secondary strategy for satellite males. Between two and sixteen males father the majority of egg clutches so it is predicted that this behaviour is a normal part of successful reproduction in frog populations across Europe.

Arrested development

Although humans tend to assume that once the developmental process has commenced it cannot be paused and re-started (seeing as this is how it works in our own bodies), this couldn't be further from the truth. Many organisms undergo a temporary halt in the developmental process, aptly termed 'diapause'. Diapausing forms are capable of withstanding a much wider variety of environmental conditions than are active forms. Arrested forms can successfully weather extreme temperatures, droughts or floods, or exist in tiny interstitial spaces that would be impossible for actively developing embryos.

Diapause is common in the invertebrate world for a variety of reasons, although one of the biggest factors at play is that of the environment. The once-popular household aquarium pets 'sea-monkeys' are the developmentally arrested eggs of brine shrimp (*Artemia* species), which resurrect once they are placed in water. Diapaused eggs of crustaceans can become extremely concentrated in some drought-stricken environments, reaching concentrations of up to one million per square metre of ground. It's an extremely useful developmental strategy to ensure that larvae will only develop under environmental conditions that provide the best chance of survival.

Other arthropods utilize the power of diapause in a more predictable fashion. For example, many butterflies and other lepidopterans have bivoltine life cycles, which means that they have two broods of offspring per year. Those living in areas that experience changing seasons will inevitably experience different selection pressures at different times of year. Embryos of the green-veined white butterflies (*Pieris napi*) in Sweden have two distinct developmental trajectories. Those in the diapause generation spend the winter as pupae, and then continue to develop once temperatures warm up. Those in the direct developing generation generally do not experience a phase of diapause because their entire life cycle happens while temperatures are warm, making a diapause stage unnecessary.

This kind of variation in life cycle according to seasonality is an extremely important (and common) strategy. Proper timing of life history events is critical to survival, and many species have evolved a large degree of plasticity in their developmental processes in order to cope with varying conditions. Bivoltine and multivoltine (many broods per year) species will experience a range of conditions throughout the year that require a variety of developmental strategies. In some cases, individuals may extend their diapause for one or more years depending on the conditions. Some species experience an increased mortality or reduced fecundity subsequent to a prolonged dormancy, but others do not. Yucca moths (*Tegeticula* species) can successfully eclose (develop) after more than seventeen years in diapause, and it's postulated that other species may be able to withstand decades, or even centuries in suspended animation – awakening only once environmental conditions are sufficient. It is like the Snow White in the insect world, only instead of a kiss from Prince Charming embryos require heat or moisture to reanimate.

Although such physical factors of the environment (i.e. temperature, moisture) are a major factor in the evolution of diapause of many species, there are other factors that potentially play a role. Predation pressure can shape the behaviour of prey species in numerous ways, including whether or not a prey species engages in diapause. Arrested embryos may be more resistant to predators because they can exist in tiny refugia that are inaccessible to them, and this is an important strategy for prey when predation pressure is high. Although it's effective at deterring predators, diapausing is not without immediate costs in terms of biological fitness, so it's not a 'decision' to be taken lightly by prey species. Reckless diapause can lead to competitive exclusion by other prey species that are successful in fending off the predators, so organisms are expected to utilize this strategy only during times of extreme stress. Indeed, water fleas (*Daphnia magna*) are more likely to develop diapausing eggs when in the presence of predatory fish that have been fed on conspecifics (members of the same species – as opposed to experimental trials where fish were fed with other foodstuffs) – indicating that the threat of predation is immediate and high. Larger, more conspicuous planktonic organisms are expected to utilize a diapausing strategy to avoid predation more often than are smaller ones, and this trend is demonstrated in several species of freshwater copepods (another small crustacean). The higher the risk, the more likely a species is to use diapause as an anti-predator strategy.

Invertebrate diapause is such a fixed part of many ecosystems that it has allowed for the fostering and coevolution of other relationships. For example, parasites that infect diapausing organisms have evolved mechanisms to cope with the fact that their host may enter into a state of suspended animation for long periods of time. Freshwater crustaceans are often the secondary hosts for microsporidian (unicellular free-swimming) parasites of fish. This means that infected individuals transfer the parasite to their primary fish hosts once they have been eaten. What's a parasite going to do if their mode of transmission into their main host decides to take a long winter's nap? It's not a flippant question because if the parasite remains virulent inside a diapausing embryo, the embryo will most certainly perish (thus killing the parasite as well). Indeed, several parasites that infect diapausing organisms enter into a diapausing state themselves until their host reanimates. The advent of diapause may actually help to regulate the spread of parasites in aquatic ecosystems. Copepods (subclass Copepoda) are the secondary host for several tapeworm parasites of fish, and their yearly diapause coincides with a seasonal immunity in the fish that tapeworm infects. In other words, the parasites safely 'bide their time' in their diapausing secondary hosts during a phase when it wouldn't be possible for them to infect the fish anyway. Talk about a sophisticated life cycle!

Speaking of sophistication, the diapausing stage of several crustaceans can also be advantageous when it comes to invading novel habitats. It seems counterintuitive on the surface (how can an inanimate embryo invade a new habitat?), but it functions in the same way that plants spread their seeds: through secondary distributors. Spiny water fleas (*Bythotrephes longimanus*) have managed to invade all the great lakes with the help of their diapausing life cycle and some unsuspecting vectors. The adult stage of the water fleas is susceptible to predation by large fish, but those adults contain diapausing eggs with thick outer shells that pass directly through the digestive system of their fish predators. The fish carry out the important work of spreading the water fleas between deep and near-shore environments, effectively mixing the gene pool so as to ensure continued development of healthy water flea populations. Fishermen who do not properly clean their lines or boats are also important vectors of spiny-flea eggs, and the use of baitfish transferred between lakes inevitably leads to further transfer. Despite its initial evolution to cope with environmental stochasticity (unpredictability), the advent of diapause in spiny water fleas has facilitated invasion into novel environments, with a little help from the irresponsible *Homo sapiens*.

Although diapause is a common part of invertebrate life histories, it's not a phenomenon limited to the spineless. Oviparous (egg-laying) vertebrates are also capable of embryonic diapause, again most often in response to extreme environmental conditions. For example, the snake-necked turtle (*Chelodina rigosa*) experiences diverse patterns of wet and dry throughout the year. These organisms inhabit wetland environments and swamps in the tropics of Northern Australia, and their activity is strictly punctuated by the dry season, where the swamps completely desiccate and adults survive by burying themselves deep beneath the ground. Females lay eggs either underwater or in water-saturated soil, and specialized modifications of the outer membrane of the eggs allow them to diapause for varied lengths of time according to ambient conditions. Eggs are still viable after six months of inundation in water (although they experience optimal survival at six weeks). The cue for eggs to exit diapause and hatch occurs when their immediate environment dries out enough for oxygen to reach them. However, the exact timing of dry down is highly unpredictable, which is essentially the reason behind the evolution of such developmental plasticity. In this species, development can be successful over a wide range of hydration regimes, which is clearly essential in the extreme and unpredictable climates of Northern Australia.

Several species of fish are also able to produce diapausing embryos. Perhaps the most specialized of these is the northern plains killifish (former genus *Rivulus*), which inhabits temporary ponds in Africa and South America. Since their main habitat is ephemeral, it's critical for them to be able to arrest development during periods of desiccation. Dry-outs inevitably kill all adults and juveniles that are present in a pond, but the arrested embryos remain intact and will animate upon re-wetting the next rainy season. Interestingly, killifish diapause can occur at three different stages. Diapause I occurs at the blastocyst stage (an extremely early stage of development prior to embryo implantation); II occurs at a later stage but prior to organ development; and III is the final, fully developed embryonic form, just prior to hatching. Diapause at stage III is obligate, meaning that it always occurs here regardless of the ambient conditions. Stages I and II are facultative, which means that there is flexible control over whether the embryos diapause according to environmental variables.

Both facultative and obligate diapause are also exhibited in several species of mammals. As with the examples from the invertebrate world, its primary function in mammals is to uncouple impregnation from birth (partuition) so as to ensure that both processes occur at the most opportune times. Again, when species inhabit unpredictable or extreme environments – such as those that inhabit Northern latitudes, it makes the most biological sense to be able to have control over timing of birth. For example, female brown bears (*Ursus arctos*) are able to vary the birth date of their cubs in response to their accumulated fat stores. Females in superior condition tend to give birth earlier in the season than those in poorer health. Several other mammalian species in the Arctic experience a fine-tuned control over the timing of their births so as to have their newborns during the most favourable time of year for finding food, being mobile and avoiding predators. Because these Northern animals cannot guarantee when they will find a sexual partner (see 'Give the Penis a Bone'), being able to 'store' embryos is critically important. Diapause in mammals almost inevitably occurs at the blastocyst stage, and is characterized by a temporary delay of embryonic implantation into the uterus.

Control of diapause in mammals lies with impregnated females, and more specifically with the chemical environment of the uterus. Facultative diapause (where its occurrence depends on physiological cues) is common in mice and rats (superfamily Muroidea). Here, females ovulate immediately following birth of their babies, allowing them to mate and create another embryo. If there is any kind of metabolic stress in the female, the newly formed embryo will enter into a state of diapause. These unanimated blastocysts are stored within the female's reproductive tract, and cellular processes within them allow for vital nutrient recycling to occur. Other mammals, such as the carnivores, undergo obligate diapause, where arrested development occurs each time an embryo is formed. Obligate diapause is often regulated by photoperiod (the length of daylight an animal experiences); in both skunks (family Mephitidae) and mink (subfamily Mustelinae) the increase in daylight after the vernal equinox is the stimulus for its termination. The change in photoperiod leads to a surge of prolactin in the uterus, which causes the embryo to implant.

Marsupials, such as kangaroos and wallabies, exhibit a combination of both facultative and obligate diapause. For example, female tammar wallabies (*Macropus eugenii*) give birth 26.5 days after the winter solstice. They experience postpartum ovulation (same as the female rodents) and so just after giving birth, they are immediately pregnant again; however, the newly formed embryo proceeds into a state of diapause (facultative). If the newborn wallaby dies, the embryo immediately comes out of diapause and the female will undergo gestation and birth again. However, if the first newborn develops normally, the diapaused embryo will remain in suspended animation until the following winter solstice (obligate).

Although diapause is currently observed in at least a few species within all mammalian orders (except primates), the bulk of mammals do not experience it (although this is debatable because the advent of diapause in internally gestating organisms is notoriously difficult to observe). It has been historically concluded that diapause is a phenomenon that evolved independently, several times only within certain mammalian groups (i.e. only those species where it actively occurs). However, recent work has challenged this notion by suggesting that embryonic diapause is an ancient characteristic that evolved only once in mammals, and although many species do not exhibit it, they maintain the ability to undertake it.

What happens if you take an embryo from a non-diapausing species, and place it into a diapausing uterine environment? Biologists answered this exact question by transplanting sheep (*Ovis aries*) blastocysts into mice (*Mus musculus*) uteri. The sheep embryos promptly entered into diapause, indicating that the ability to do so remains physiologically intact. The embryos were then transplanted back into the sheep where they developed into normal lambs, indicating that the process of diapause had no ill effects on them. Another study reciprocally transplanted embryos between non-diapausing ferrets and diapausing mink (both in the family Mustelidae). The former entered diapause, and the latter emerged from it. So if the incidence of diapause is truly conserved across all mammals (and so far experimental evidence supports this conclusion), then it would make sense for it to be possible in primates as well. Biologists speculate that if embryonic diapause is indeed possible for primates, and specifically humans, this could revolutionize the science of reproductive biology. Clearly it's not possible to test this directly on primates, but the idea is intriguing and merits further thought. After all, there is clear variation in the gestation length of human foetuses, and a simple measure of hormonal levels could clear up whether human females are diapausing or implanting. Inquiring minds want to know!

Tired old ladies and dirty old men

Although we humans (females especially) have developed a multitude of strategies to attempt to stave off the inevitable, the ugly truth about living is that each and every day you are one day closer to dying. There are of course innumerable ways a person or animal could meet their death, but for those of us lucky enough to escape from predators, accidents, illness or random horrible things the process of senescence will be the major contributor to our ultimate demise. Senescence, essentially the process of ageing, is unavoidable reality – perhaps more so for humans than any other animal because we have developed so many technologies and medical interventions that keep us alive longer. For decades it was widely assumed that animals (other than domesticated ones) didn't live long enough to undergo senescence. Challenges like predation, competition, resource availability or flood/fire events were thought to result in the death of most animals before they ever had a chance to get old. However, more recent advances in the studies of ageing have confirmed that most animals do indeed undergo senescence – especially mammals and birds.

Actuarial senescence pertains to the ageing of the body; in humans we can think of this as general health issues in the elderly – wrinkles, weaker bones, lower energy and loss of muscle mass. Evolutionary explanations for the advent of actuarial senescence are based on the fact that natural selection is not as strong in older organisms. Deleterious mutations with late-acting effects are therefore able to accumulate in elderly creatures, making them weaker and more prone to fall victim to illness or injury. The other kind of senescence is *reproductive*, and it pertains to the fact that in several animals there is a marked decline in reproductive function with age. Human females are nearly unique in the animal kingdom in that we experience a complete and irreversible cessation of our reproductive capabilities at around the midpoint of our lifespan. Most female animals (including other primates) retain the ability to produce offspring, though the quality and quantity are affected by maternal age. I am going to discuss the evolution of menopause in humans and whales in a subsequent section, for now I'm going to focus on the diverse ways that animals age and their strategies to deal with it.

The theory of antagonistic pleiotropy (what a mouthful!) explains a high proportion of the occurrence of reproductive senescence in diverse animals. It essentially states that organisms exhibiting a large investment in reproductive efforts *early* in their lifespans are doing so at the expense of later-life reproductive prospects. If overall reproductive output is increased by engaging in a greater amount of reproduction earlier (when selection pressure is stronger), then it is possible that it's worth trading off against later reproductive success. Related to the concept of antagonistic pleiotropy is the notion of 'disposable soma', which is a fancy way of saying that if you're spending all of your time and energy on reproduction you won't have enough energy left for maintenance of your somatic (non-reproductive) tissues.

So in addition to experiencing reproductive senescence, chances are you're experiencing actuarial (non-reproductive) senescence as well. They key thing for any individual is to weigh the costs and benefits of the present versus future prospects for reproduction, which isn't always easy. Not to mention that environmental or population-level complications could intervene at any time and make the situation more complicated. Despite the potential for complicating factors on an individual level, there are several trends at the species level that indicate that there is more than one way to tackle the inevitable process of ageing.

For example, both sexes of common guillemots (*Uria aalge*) (a long-lived seabird) experience a significantly reduced breeding success in the last three years of their lives. In accordance with antagonistic pleiotropy and disposable soma, individuals that engage in high levels of reproductive output at early life stages experience a greater level of senescence earlier than those with lower levels of early reproduction. Lifespan in guillemots is directly related to reproductive output: the earlier you start making babies the earlier you are likely to die. Not entirely surprising, but a little depressing all the same. A similar trend is evident in red squirrels, although in these mammals females have a greater chance of experiencing reproductive senescence. Older mothers produce offspring with lower rates of survival, which is attributed to the fact that they themselves are of lower quality. Although females retain the ability to produce offspring to an old age and may produce as many offspring as younger mothers, their babies are of lower quality and have a lower rate of survival. One might assume that female mammals would experience a greater level of reproductive senescence based on the fact that they are gestating and lactating for their offspring – both of which entail massive energetic and physiological costs. However, there are times when the costs of reproduction are high for males as well, especially those that have to engage in intense competition in order to be reproductively successful.

For example, male red deer (*Cervus elaphus*) that have large harems of females in their early lives experience high levels of senescence later on. Despite the fact that males do not directly contribute to parental care, gestation or lactation, there is intense sexual competition between males for the affections of the ladies. Red deer are highly polygynous, males aspire to achieve large harems of females during the annual rut. They do this through development of huge antlers, which are the only mammalian structures that are regenerated repeatedly every year. Antlers can comprise up to 35 per cent of a male's body weight, and they grow faster than any other mammalian tissue. A substantial investment in these expensive sexual structures will improve a male's chances of reproduction, but it will also lead to an increased reproductive senescence (in terms of harem size and rut duration) earlier in adulthood. It follows that males have higher reproductive costs in polygamous scenarios than in monogamous ones, and we expect to see this reflected in the levels of reproductive senescence. However, even monogamous males still have to impress a female before they get a chance to mate with her, and older males often have a harder time creating and maintaining expensive sexual signals.

Male blue-footed boobies (*Sula nebouxii*) are selected by females based primarily on the level of vibrant blue-green coloration of their feet (see 'Set Yourself Up for Success'). Signal quality inevitably declines with age, because foot colour is based on levels of antioxidant pigments that must be continuously supplied. Elderly males experience a build-up of free radicals and oxidative damage, which leads to the activation of their immune systems. This is a common problem in any elderly animal, but for the booby, the antioxidant pigments that would normally provide them with brilliant blue feet are otherwise tied up with fighting free radicals. This results in their feet turning a duller colour, which means they achieve a lower level of female attention. Bring on the golf clubs.

A similar thing happens with male yellow mealworm beetles (*Tenebrio molitor*), which experience age-related decreases in reproduction. First, they experience pre-meiotic sperm senescence, which is along similar lines to the physiological issues experienced by the blue-footed boobies in the example above. Essentially, the ageing genome is more likely to experience mutations and produce sperm that are of a lower quality. Secondly, they experience post-meiotic sperm senescence when sperm are then subjected to a storage period within a female's spermatheca. Sperm that have experienced pre-meiotic senescence are particularly prone to post-meiotic senescence, which is bad news for male beetles that are trying mate. Females actively avoid having to procreate with these 'dirty old men', and when they are experimentally forced to do so, the offspring are of lower body weight and poorer quality.

A behaviour called 'spermatophore guarding' generally takes place in several beetle species. This is when post-coital males and females remain in close contact with one another to allow the sperm enough time to transfer to her storage organ. Males can aggressively defend 'their' females, and females that want to ensure his paternity remain in close contact in order to limit harassment by others. However, when mated to an elderly male, females drastically decrease their participation in spermatophore guarding, thereby increasing the opportunity for sperm competition (i.e. seeking additional copulations). Old males actually increase their efforts at spermatophore guarding subsequent to copulating with a female, which is likely the best (and only) thing they can do.

The environment can also play a key role in the impact of ageing. If resources are always plentiful and abundant, we expect organisms to 'live life to its fullest' and exhibit increased measures of sexual reproduction. During periods of resource shortage we expect the opposite to be true, which is bad news for reproductive output but possibly good news for overall senescence. Experimental work done on hermaphroditic snails highlights the costs and benefits of reproducing in varied resource scenarios with respect to reproductive senescence. Groups provided with ample, nutritious food experience greater reproductive senescence than those fed a low-nutrition diet. Indeed many organisms have been observed to delay the onset of sexual reproduction when resource levels are low – which may indirectly decrease the level of reproductive senescence experienced in later life. In the case of the snails, being given a massive amount of high-quality food leads to surges in sexual reproduction and a faster onset of senescence.

While most animals experience an increase in fecundity up to a certain age, and then a decrease through to their deaths, there is another strategy that can serve to increase reproductive output at the end of their reproductive lives. As we have already seen, terminal investment describes a scenario where ageing or sick animals increase their allocation to reproduction so as to make the most of their energies while they are still able (see 'Sex or Sick'). This scenario has been observed in the wandering albatross (*Diomedea exulans*), where longitudinal data sets document the reproductive investment of specific individuals over a number of breeding seasons. Albatross are extremely long-lived birds (over fifty years in some cases) that are socially monogamous and produce a single-egg clutch each year. Males and females both exhibit a high level of parental effort, and both experience actuarial and reproductive senescence. Following their peak reproductive age, there are senescent declines in reproductive performance followed by a striking increase in breeding success and key parental investment in the terminal-breeding attempt. In other words, when survival probability and prospects for future reproduction fall below some kind of threshold, the albatross 'go for broke' and put all their remaining energy into their final chance at increasing their biological fitness. Certainly a tantalizing prospect, although biologists have yet to determine the mechanisms by which it happens.

Cooperatively breeding mammals provide another interesting system in which to examine the effects of reproductive senescence, because reproductive costs vary according to one's place in the social hierarchy. For example, meerkats (*Suricata suricatta*) live in large social groups of both sexes. As we have seen, one alpha female, who is behaviourally dominant over all others, is the only one who undertakes reproduction. She may go so far as to kill the offspring of subordinates should they be so bold as to create any. Submissive females rarely breed, instead taking on a helping role for the offspring of the dominant – who produces between one and four litters of up to seven pups per year. The pups are cared for by many group members, and may even be fed by subordinate females through allolactation, although this is rare. After approximately three weeks, pups emerge from the underground den and begin to forage (with help) with the group. Despite the fact that meerkat alpha females experience a great deal of reproductive support from conspecifics, they still experience reproductive senescence. In fact it occurs relatively early in dominant females, although the specific mechanisms for it are unknown. Intuitively, you want to conclude that the fact that a female mammal is gestating and lactating so many offspring takes a negative toll on her ageing body. In fact, that much seems an obvious conclusion. However, we are then faced with the example of African mole rats (family Bathyergidae), whose reproductive biology flies directly in the face of this idea.

Similar to the meerkats above, African mole rats are cooperative breeders with very low mortality and unusually high longevity. They live in underground dens in multi-generational family groups, and only a few key individuals engage in sexual reproduction. Most of the offspring of the alpha (reproductive) females remain in the den as non-reproductive helpers for several years. Helping females have fully functional ovaries; however, they do not ovulate in the presence of the dominant. Remarkably, breeding females seem to age less than their non-breeding counterparts. For some reason, the females that are doing all of the reproducing realize a longer life and fewer pitfalls of growing old. In what seems to be a reversal of the classic trade-off between somatic maintenance and sexual reproduction, this same trend has been shown for giant (*Spalax giganteus*), Ansell's (*Fukomys anselli*) and Damaraland (*Fukomys damarensis*) mole rats as well. And we're not talking about small increases in longevity either: reproductive Ansell's mole rat males and females live roughly *twice* as long as their non-breeding counterparts (more than twenty years for breeders). This system is parallel to what is observed in many species of eusocial insects (like ants, termites and bees) where reproductive queens have a much longer lifespan than non-reproductive helpers.

Due to the high level of relatedness in mole rat groups, the pronounced social structure may be advantageous from a kin-selection perspective and also to decrease the incidence of incest, seeing as many brothers and sisters are living in the same space. So what could be the reason behind the increased longevity in reproductive individuals? Is it some factor of the provisioning that somehow relaxes the ageing process? Eusocial ant queens realize a hundred-fold increase in lifespan in comparison to solitary species, which is directly attributed to the provisioning she receives from workers. Who wouldn't survive longer when you hardly have to lift a finger, and every need is met by your trusty colony? Mole rat queens don't have to spend any time foraging or maintaining colony structure, and they are aided greatly in terms of care of newborn offspring. So should we assume a decrease in the lifespan of mole rat helpers due to a high stress level? Researchers do not suspect this, because mole rats do not experience competition for mating partners, hardly any competition for food, and generally have a very low level of aggression. On the contrary, individuals are often seen engaged in sociopositive behaviours like grooming and communal sleeping.

Interestingly, when a helping (non-reproductive) female is removed from the colony and placed with a male, she begins to ovulate in addition to experiencing a number of other physiological effects including an increase in body length and brain volume, and pituitary sensitivity. It's thought that these changes are associated with decreases in the ageing process. Tissues of reproductive females contain significantly lower levels of oxidative damage than those of their non-reproductive counterparts, which seems to point directly at a process aimed at combating oxidative stress. As described in many of the above examples, the level of damage from free radicals (oxidation) is a major contributor to the ageing process in many animals. Have female reproductive mole rats evolved a way to directly combat this? It seems they have, although the exact mechanism by which they are accomplishing it has yet to be determined. I wonder if some pharmaceutical company has a patent on reproductive mole rats . . .

Grandmamas

In the previous section we discussed the occurrence of reproductive senescence, or the natural decline in reproductive output
and capability with increasing age. This kind of sexual slowdown
is common across the animal kingdom; however, humans and
toothed whales represent an extremely specialized form of reproductive senescence. Here, females experience a complete and
irreversible cessation of reproductive capabilities (menopause) at
a mid-point in their life cycles, which means that they exist in a
state of biological irrelevance for a good portion of their lives.
Human females rarely give birth after age forty-five following a
fertility decline lasting two decades. Similarly, short-finned pilot
whales and killer whales stop reproducing by ages thirty-six and
forty-eight respectively, but live to the ripe old ages of sixty-five
and ninety. Why would such a strategy evolve? After all, males
retain the ability to procreate until the end of their life cycles
(albeit with reduced success for both physical and social reasons).
Why haven't females done the same?

There are a few main hypotheses for the existence of menopause, and they are not mutually exclusive. In humans at least, the
advents of industrialization and modern medicine have extended

our lifespans far beyond those of our closest primate relatives. Indeed, chimpanzee (*Pan troglodytes*) (one of our closest relatives) females *can* experience reproductive senescence to the point of complete cessation; however, they rarely live long enough. The data points to a clear decrease in reproductive output of female chimpanzees followed by a complete cessation in their fourth decade (if they are still alive). This is because in chimps and most other primates, actuarial (somatic) senescence and reproductive senescence occur at comparable rates, whereas in humans there is a distinct uncoupling of the two (with reproductive senescence occurring at a much faster rate than actuarial). It's possible that ancestral humans experienced a similar reproductive pattern to the great apes, where reproductive senescence was followed quickly by death. Since humans in Western societies have increased our (somatic) lifespans to such a great degree, the advent of menopause has been 'uncovered' relative to other primates. This suggests that complete reproductive senescence may be a by-product of our longer lifespans and the fact that reproductive lifespan has not evolved to 'keep up' with that of the somatic lifespan.

Some researchers suggest that this is because there is a finite 'shelf' life to a human egg, and since human females are born with all the eggs that they will ever utilize, extending their reproductive lifespan without introducing donor eggs is impossible. However, other mammalian females (such as whales and elephants) continue to reproduce into their sixth decade, suggesting that it's not impossible to increase the lifespan of a mammalian egg. In addition, the effects of modernization cannot completely explain the large post-reproductive lifespan of human females, because data from modern-day hunter-gatherer societies show that these ladies also experience substantially long post-reproductive lives.

The second main hypothesis for the advent of reproductive cessation is an adaptive one. Picture it: as ancestral humans were expanding their populations across African savannahs in the Pleo-Pleistocene era (between 1.7 and 1.9 million years ago), the increasingly arid conditions meant that food sources were changing. Juveniles that were once able to gather food were no longer able to forage for roots and tubers located deep beneath the ground because the strength of their small arms was insufficient to dig through the hard earth. This resulted in a conundrum for ancient mothers: remain on a nomadic path in order to follow food sources that their young children could manage, or stay in a constant area where food is tough to find, and provision the children for longer periods of time. If they chose the latter, this would require a greater expenditure of their maternal energy on their current offspring, leaving them with limited chances of further reproductive success. Mothers could not overburden themselves with extra offspring because the costs of their own death would be great – juvenile humans are unlikely to survive without direct parental provisioning, so an over-taxed mother could lose her entire brood if she fell ill or died.

This scenario presented a valuable opportunity for older females (already experiencing natural reproductive senescence) to increase their own inclusive fitness. You see, if an elder (grandmother) stepped in to help provision her grandchildren it would be biologically beneficial for both her daughter (who could have more offspring at a faster pace without having to sacrifice the well-being of current offspring) as well as herself (through the increased survival of her grandchildren and the increased prospect of further grandchildren). More vigorous grandmothers could provide a higher level of care, which would be under a high level of selection through the direct fitness benefits conferred. In other words, selection may actually have favoured the earlier occurrence of complete reproductive senescence because grandmothers could gain more net biological fitness through helping than through increased child rearing of their own. Gestating, birthing and providing milk for a human infant represents a huge energetic and physiological cost to females. To put it simply: the massive costs of child rearing in our species are better faced by a younger body than an older one. However, by provisioning their grandchildren, older females can continue to increase their own biological worth. So in this unique situation (aptly termed the 'grandmother hypothesis') the end of fertility is *not* the end of reproduction.

Tellingly, there has been no mention of grand*fathers*. Do grandfathers contribute to their collective fitness in a parallel way to their female counterparts? So far all the data suggest that they do not. This is due to the ever-important notion of paternity certainty for both fathers and grandfathers. A mother is 100 per cent certain that she is the parent of her offspring (yes, the memories of childbirth would be enough to solidify this in any female primate's brain), whereas a father does not have this same security. Similarly, a grandmother can be certain that she is the grandmother of her female offspring's children. She knows who her daughter is, and the babies that make their way into the world through her daughter's birth canal are most certainly her grandchildren. The same level of certainty cannot be applied to fathers or grandfathers, so their investment in children and grandchildren is naturally lower. Following from the notion of paternal uncertainty is the fact that the bulk of grandmother provisioning is observed along maternal lines. However, data collected from modern and historical traditionally living human populations (without Western comforts or medicine) show a noteworthy trend with respect to maternal versus paternal grand-mothering. The *X-linked grandmother hypothesis* is an extension of the grandmother hypothesis that pertains to relatedness between grandmothers and their grandchildren in their X chromosomes.

Maternal grandmothers have 25 per cent relatedness to both their male and female grandchildren by virtue of the fact that they contribute one X chromosome to their daughter (who receives one from her father as well), and then one of these X chromosomes is passed on to both her male and female offspring. For *paternal* grandmothers the story is a little different. Male children of a given mother will have inherited the X chromosome directly from her (the father's contribution is the Y chromosome). Daughters of this male (i.e. paternal granddaughters) will have this same X-chromosome since that's the only one the male has. This means that paternal grandmothers have 50 per cent relatedness to their granddaughters. Sons of this male (i.e. paternal grandsons) will inherit his Y chromosome (not his X), meaning that a paternal grandmother has 0 per cent relatedness to her grandsons when it comes to the X chromosome.

How might we expect these differences in relatedness to play out in the real world? Do grandmothers play favourites based on their relatedness in a maternal or paternal grandmother context? It appears that they do. A large dataset comprising population statistics for seven traditional human populations varying in both geographical (Japan, Germany, England, Ethiopia, The Gambia, Malawi and Canada) and temporal (seventeenth to twenty-first century) space showed that boys survived better in the presence of maternal grandmothers, and that paternal grandmothers have a more beneficial effect on girls. In some populations paternal grandmothers actually had detrimental effects on grandsons.

The exact mechanisms for this type of grandmother bias in provisioning are not fully understood, although it could see its basis in pheromonal cues or be based on physical likeness between grandmothers and grandchildren. I would imagine that these effects are all but absent in modern Western societies, seeing as the notion of paternity certainty is virtually a moot point in our largely monogamous populations. Any potential uncertainty can easily be remedied by a quick DNA test, meaning that all grandparents have the opportunity to know their exact relatedness. However, just because a grand-daddy knows who his grandchildren are, does not necessarily mean that he contributes anything other than the occasional ice-cream cone or pony ride.

Could there be something akin to the grandmother hypothesis for grandfathers? Do grandfathers in more modern-living monogamous societies contribute in a meaningful way to the well-being of their grandchildren? According to a study using a multi-generational dataset of pre-industrial Finnish grandfathers, they do not. The work examined whether the presence of grandfathers conferred any positive effects on age at first reproduction, birth intervals, offspring lifetime fecundity and reproductive success, and in no parameter was there any significant correlation. Why then do human males have such an extended lifespan (akin to their female counterparts)? It's thought that long life in males arose through the benefit of continued (successful) mating throughout their lifespans, or as the unselected byproduct of selection of genes for longevity in females. To me, the latter seems more realistic. The role of modern grandfathers may simply be confined to enjoying the grandchildren from time to time when their golf schedules permit.

So far our discussion has focused on primates, and specifically humans, because we are unique in our expression of reproductive senescence early in our lives. Unique, but not exclusively so. As I mentioned above, there are some remote instances where female primates do live long enough to be of help to their reproductive daughters. In Japanese macaques (*Macaca fuscata*), the reproductive pattern of females varies according to whether their mothers are around to help. Those that have mothers present (whether or not those mothers are still reproducing themselves) begin reproducing successfully at an earlier age than females who do not. In addition, if her mother is both present and post-reproductive, a daughter will have more babies in a shorter timespan. These results show an analogous situation to the ancestral humans on the African savannahs millions of years ago – when there is another set of trustworthy (related) helping hands, this frees a female to increase her number of offspring without direct cost to the ones she already has.

The other group of animals where we see a prevalence of menopause with females experiencing a significant portion of their life in a post-reproductive state is in the toothed whales (suborder Odontoceti). In fact, killer whales have the longest post-reproductive lifespan of all nonhuman animals. What kinds of benefits do these cetacean grandmothers confer to their offspring and grandchildren? The social structure in cetaceans is quite different from that of humans. There are cooperatively breeding troops where both males and females remain close to their home group, which means that a female's relatedness to group members increases as she ages – setting the scene for her reproductive senescence. In other cooperatively breeding mammals like meerkats (*Suricata suricatta*) and hyenas, younger members of the group give up reproduction in order to supplement the reproductive efforts of their parents. They remain in their natal groups and assist in offspring care, which has the ultimate result of increasing the biological fitness of their parents and themselves. In menopausal species like killer whales (*Orchinus orca*), we see a complete reversal of this pattern. Older females give up their reproductive capacity to assist in the breeding efforts of their daughters. Interestingly, there is an asymmetry in the benefits that a grandmother realizes by helping her sons versus her daughters.

For male whales, mating occurs outside of the matriline. Essentially, he will leave the group to find a mate, impregnate her and then return home to mother without contributing anything further to the existence of his offspring. The opposite is true for daughters, who leave the group temporarily in order to get pregnant, and then return to the group to birth and raise their offspring. The addition of more young offspring to the group increases the workload and the level of competition for all group members, so it's actually more beneficial for older mothers to contribute to the well-being of their sons over their daughters. In other words, a greater level of biological fitness can be conferred by ensuring that her sons are able to continue to propagate her genes while she doesn't have to do any additional work for it. In line with this, males older than thirty years experience an 8.3-fold increase in death rate in the year following their mother's death, whereas the rate for females older than thirty is 2.1-fold. It seems mother offers a good deal of favouritism when it comes to the protection of her adult offspring from antagonistic interactions or providing help during foraging. It is extremely beneficial for old mothers to prolong the reproductive lives of their adult sons, because for male killer whales reproductive success increases with age.

The scenario for killer whales is almost the opposite of what we observe in primates when it comes to the notion of grandmothering. The evolution of primate menopause tends to fall along benefits to those in matrilines and particularly from mothers to daughters; whereas in killer whales the evolution of menopause sees a greater female expenditure on the survival of her sons, which indirectly contributes to his reproductive success outside of any contact with the menopausal grandma. These are two very different strategies for the evolution of reproductive cessation in these mammalian groups.

POSTSCRIPT

It's been an amazing, diverse journey through the trials and tribulations of the sexual process in the animal kingdom. My main goal in writing this book was to enlighten the reader to the fact that sex is a part of almost all decisions made by animals, and it contributes to virtually every facet of social organization and community formation. Perhaps you decided to read this book with a completely different viewpoint about sex, and perhaps the sheer diversity, complexity and far-reaching biological themes have bent your mind in new directions. 'Animal Sex Biologist' is a title that's associated with a solid understanding of many aspects of biological and sociological science – it's not some flippant term that pigeonholes me into exclusivity on funny penises or horrifying sexual encounters. It's a title I've earned, and one I'm extremely proud to bear.

There are a few main themes from the book that merit further mention since they are central to the study of sex:

Gamete Value: Time and time again I find myself attributing a structure, behaviour, process or pattern to the drastically different strategies of males versus females. The truth of the matter is that most sex is antagonistic. When the investment for one sex (females) is so high in comparison to the other, it can only follow that conflict will result. The low value and massive abundance of sperm, contrasted to the high value and finite quantity of eggs sets the scene for the majority of sexual strife that we've covered in this book. While antagonism isn't ubiquitous, it's of enormous importance in the world of sex.

Evolution at Multiple Scales: This isn't an easy notion to grapple with because humans tend to think of adult animals as the unit upon which evolution acts. Certainly natural selection does manifest at the adult stage of animals, but also at any and all stages of development. Juvenile individuals (especially those with low levels of parental investment) experience a different suite of selective pressures than do adults, and it follows that only the 'fittest' infants will ever grow up to become them. In addition to the fact that selective pressure is alive and well at all stages of development, there is the fact that it exists at levels other than that of the whole organism. Particularly import-ant to the topic of sex is evolution at the cellular level: that of sperm and egg. As we saw in several sections of this book, there is tremendous diversity in the quality of individual sperm, as evolution will favour those that are most able to achieve fer-tilization success. So as well as the selective pressures affecting an individual at any given time (with respect to sex or survival), the forces of evolution are at work on the level of the gamete (not to mention any number of other scales like tissues, organs and organ systems).

Homosexuality: Same-sex copulations and relationships are ubiq-uitous in the animal kingdom. If anyone attempts to establish that this is not the case, please have them read this book. If that doesn't work, please refer them back to this book.

Hermaphrodites and Parasites: Traditional thinking about the sexual process fails to take into account the fact that alternative systems exist. Although they are completely different from each other, hermaphroditism and parasitism present biologists with decidedly complicated scenarios that fall far outside the realm of what is considered to be *normal*. These two scenarios are reminders that not all procreative sexual systems involve just a male and a female.

The study of sex in hermaphrodites is an area that I find to be fascinating, yet also incredibly daunting. These systems are among the most violent and brutal when it comes to copulation, yet at times it is almost impossible to decipher the needs of both the female and the male parts of an individual when either (or both) of them are being affected by actions of the male and female parts of the sexual partner. I wonder if hermaphrodite sex is something we will ever truly understand. It's the ultimate in sexual conflict, and the strategies by which individuals become reproductively successful are some of the greatest tales in this, and other books on the diversity of sex.

Parasitism represents another captivating situation when it comes to sex. A parasite is a 'third party' at the table with its own biological fitness to maximize, and it maintains a good deal of control over the actions of its hosts. The impacts of parasites on sexual behaviour of hosts are again difficult to decipher, but their potential is enormous. When host animals lose control over their sexual behaviours, these changes reverberate through ecosystems in complex ways. In addition, the mechanisms by which parasites mediate host activities are fine-tuned to the extent that parasites maximize their own propagation while keeping hosts alive and healthy enough to maintain sexual function.

US versus THEM: One of the things that fascinates me the most about sex is the fact that our own species does so many things so differently than the rest of the animal kingdom. Essentially, humans have taken the biology out of the sexual process. Will this lead to our ultimate demise?

I have refrained, for the most part in this book, from including sarcastic jabs about how humans get it wrong on so many levels when it comes to our basic biology. Essentially, the enormous complexity upon which this book is based doesn't pertain all that much to members of our own species. Perhaps our biggest biological blunder is the fact that we completely ignore the notion of maximizing our fitness by utilizing contraception. Don't get me wrong: I'm happy to NOT bear all the children that my body would allow – it gives me the opportunity to have outside interests, a career, to write this book! However, on a biological level I'm doing my genetic blueprints a disservice. Those that have large numbers of children will have a greater impact on the evolutionary future of the human race than I will. It's just a fact.

In addition to our near-ubiquitous use of contraception, humans allow most of our natural biology to be concealed using any number of products (makeup, hair colour, deodorant, shampoo, clothing and jewellery). For other animals out there, good genes are responsible for health and beauty; however, for the human animal this is not the case. Less than optimal genetic blueprints can easily come in decent-looking packages, which has the potential to confuse what should be the most important biological decisions of one's life.

The human animal of the Western world is also blissfully re-moved from the processes of hunting and gathering, such that our energy can be more directly devoted to relationships, sex and parenting. So many of the sexual strategies for procreation that we learned about in this book hinge on resource availability, whereas for our species in the Western world it's a moot point. But is our overabundance of time and energy for childcare the best thing for them? I'd like to run that question past all parents who still have children over the age of thirty living at home. If humans took a parenting cue from our closest primate cousins, juveniles would largely be without direct parental care as soon as puberty hits. In addition to providing an overabundance of care to our own biological sire, we also provide care to the chil-dren of our peers (play dates, child minding), and in many cases we provide foster care or adopt unrelated juveniles. There is no direct advantage for doing these things, although we do them just the same. On a strict biological level, any parental care provided to non-genetically related offspring represents both a decrease in our own fitness and an increase to someone else's.

For me, one of the most fascinating things about human sex-uality is our adherence to sexual and social monogamy. Almost no other mammals exhibit such a strategy for valid biological reasons. Where did monogamy begin? Why did it catch on so completely in our species? While there is an abundance of research on the topic, I'm unsure as to whether there is one overriding conclusion. Although it may not be 'what all the other animals are doing', I am still a believer in the notion of a monogamous relationship. I think that this is because while our bodies may want to, our brains don't allow for it. The complex suite of emotions that accompanies any sexual relationship lends itself to having one partner (or one preferred partner) over being highly promiscuous. That's not to say that other animal partnerships don't involve emotion – it's just that our own emotions are the only ones that we understand.

When all is said and done, the lack of biology in our sex lives does make our overall lifestyle pretty comfortable and easy. While the notion of maximizing biological fitness is utterly out the window for the human animal you won't be hearing me complain about it. My house is warm and stocked with food, my bedroom is cosy and comfortable, and my IUD is doing its thing to ensure that I am through with spreading my genetic blueprints.

PREFIXES

A– : lacking, without; e.g. aflagellate = without a flagellum, aphallic = without a penis

Allo– : different, other; e.g. allosperm = sperm produced by another

Auto– : self; e.g. autosperm = own sperm

Con– : together, with, same; e.g. confamilial = in the same family, congeneric = in the same genus, conspecific = in the same species

Eu– : well, good, true; e.g. eusocial = good or high degree of social behaviour, euphallic = possessing a true penis

Hetero– : different, other (opposite of Homo); e.g. heterosexual = between different sexes, heterospecific = between different species

Homo– : same (opposite of Hetero); e.g. homosexual = between members of the same sex

Hyper– : excessive, a lot, more; e.g. hyperactive = more active than normal, hypersexual = more sex than normal (than average of other species)

Inter– : between; e.g. interspecific = between different species, international = between nations

Intra– : inside, within; e.g. intraspecific = within a species, intracellular = within cells

Para– : beside, beyond; e.g. paragenitalia = reproductive organs beyond or in addition to regular genitalia

GLOSSARY

Actuarial senescence: age-related changes of the body; the decline of bodily structure or performance with increased age.

Allohormone: a compound that is passed from one individual to another and acts to directly influence physiological responses (bypassing sensory organs).

Altricial: young that are relatively undeveloped and/or helpless when born. *See* **Precocial**.

Amplexus: a type of mating behaviour where the male grasps the female for a prolonged period of time, externally fertilizing her eggs when she releases them. Practised by many species of amphibians and horseshoe crabs.

Analogous: structures or genes with similar functions but different evolutionary origins; e.g. wings on bees and birds. *See* **Homologous**.

Andromorph: an individual that resembles the male version of a species. *See* **Gynomorph, Gynandromorph**.

Anthropogenic: human-caused, or the result of human activities.

Anthropomorphic: possessing or attributed with human characteristics, feelings or appearances.

Antrum: catch-all term for cavities in bodies. When used alone, generally refers to part of the invertebrate reproductive tract where sperm is received (whereas 'antrum follicularum' is a cavity in the epithelium that envelops oocytes in vertebrates).

Aphallic: lacking a penis. *See* **Hemiphallic**.

Apophallation: deliberate amputation of the penis in snails to separate after mating.

Atricial: born or hatched in such an undeveloped state as to require substantial parental care. *See* **Precocial**.

Baculum: an ossified structure (bone) which supports the penis; homologous to the baubellum.

Baubellum: an ossified structure (bone) found in the clitoris; homologous to the baculum.

Behavioural syndrome: a set of behavioural traits that are correlated to one another.

Biological fitness: relative reproductive success of a given individual or trait; note that 'fittest' is the one that produces the most offspring, and thus is not necessarily related to muscle mass, speed, or human measures of being physically 'fit'.

Bower: a dwelling or attractive retreat or room; biologically speaking, a structure made by males to attract the opposite sex.

Bursa copulatrix: an extension of the tail cuticle of nematode worms that functions as a copulatory structure or penis.

Chromatophores: pigmented cells or groups of cells that reflect light, thus giving living things coloration.

Cloaca: a single cavity at the end of the reproductive tract that is used both for excretion and reproduction.

Coercion: forcing another to do something through physical overpowering or threats; i.e. sexual coercion is when one partner forces the other to engage in some form of sexual activity.

Copulation: the act of sexual intercourse; also referred to as mating, sex, or fornication.

Cryptic: concealed or camouflaged; e.g. an animal may be considered cryptic if it blends in with its surroundings. *See* **Cryptic female choice**.

Cryptic female choice: female control of male insemination success through physical or chemical mechanisms.

Cuckoldry: the act of unwittingly investing parental effort in offspring that are not genetically related to the parent; can be within a species (e.g. males raising young sired by another male) or between species (raising young that are of a different species).

Cuticle: the outermost layer of a living tissue, often referring to the waxy coating of an arthropod's exoskeleton.

Dewlap: loose skin which hangs from the neck of an animal.

Diapause: a period of suspended development of an animal, often during stressful or otherwise unfavourable environmental conditions.

Duetting: singing a song with another individual; in animals, often refers to specific patterns of turn-taking by singing birds.

Ecology: the relationship of an individual to its environment, or the branch of biological science that studies such interactions.

Egg: *See* **Gamete**, **Ovum**.

Ejaculate: the primary sperm-containing fluid that is emitted by a penis during orgasm; semen; cum. Not to be confused with pre-ejaculate, which is a transparent, viscous fluid emitted during sexual arousal but prior to orgasm.

Female defence polygyny: social structure where a male protects a group of females that are already social for unrelated ecological reasons.

Gamete: a mature reproductive cell that contains half the DNA from its parent, and must unite with a gamete from the opposite sex to produce viable offspring. Male gametes are called sperm, while female gametes are called ovum or eggs. *See* **Ovum, Zygote**.

Genotype: a specific combination of gene variations; can refer to a subset of genes or the entire genome.

Gynandromorph: an individual that looks both male and female. Can be mosaic (mottled around the body) or bilateral (down the middle, with half appearing male and half appearing female). *See* **Andromorph, Gynomorph**.

Gynomorph: an individual that resembles the female version of a species. *See* **Andromorph, Gynandromorph**.

Haemocoel: the fluid-filled primary body cavity of an invertebrate.

Haemolymph: the circulatory fluid in invertebrates analogous to blood in vertebrates.

Haplodiploid: the genetic system where females develop from fertilized eggs (diploid) while males result from unfertilized ones (haploid).

Hemiphallic: possessing an underdeveloped or degenerate penis, as opposed to a fully developed one (Euphallic) or none at all (**Aphallic**).

Hermaphrodite: an individual with both male and female sex organs or genitalia.

Homologous: structures or genes arising from a common evolutionary ancestor. *See* **Analogous**.

Infanticide: the killing of an infant or young member of a species. *See* **Siblicide**.

Intromission: the insertion of penis into vagina during intercourse.

Lek: an assembly of males during the mating season to compete for reproductive access to females.

Menopause: the time in a female's life when she stops experiencing oestrus; reproductive senescence in mammals.

Monandry: having one 'husband' or male partner at a time (adj: monandrous).

Morphotype: a distinct subgroup within a population distinguished by physical; a morph.

Musth: a 'frenzied' or hyperactive state of male animals when they are reproductively active, like during the mating season (e.g. elephants).

Neotropical: of or relating to the tropical biogeographic region of Central and South America.

Oestrus: the recurring period of fertility in female mammals, commonly referred to as 'heat'.

Olfaction: detection of chemicals through nasal sensory organs; 'sense of smell'.

Operational sex ratio: the ratio of fertilizable females to sexually active males.

Oviparity: reproduction through the laying of eggs (adj: oviparity). *See* **Viviparity**.

Oviposit: to lay eggs.

Ovotestes: hermaphroditic organs which produce both sperm and eggs, as found in some snails.

Ovum: the mature female reproductive cell (gamete); also called an egg. *See* **Gamete, Zygote**.

Oxytocin: a hormone and neurotransmitter involved in pair bond formation and certain reproductive behaviours.

Paragenitalia: structures alongside the gonads that occur in some species with traumatic insemination to protect against bodily harm.

Parapatric: possessing ranges which are adjacent and either do not overlap or overlap only along a narrow border.

Paternity confusion: the act of obscuring or otherwise making it impossible to determine the sire of offspring.

Pedipalps: appendages on the heads of arachnids, used in sperm transfer.

Phalloid: penis-like.

Phenotype: the outward manifestation of an individual; how an individual looks.

Pheromone: a chemical produced and released into the environment by one individual to influence the behaviour or physiology of other members of its species.

Pigmentation: coloration or degree of coloration (pigment).

Pleiotropy: the phenomenon where one gene (or locus) controls more than one trait (or phenotype).

Polyandry: having more than one 'husband' or male partner at a time (adj: polyandrous).

Polygamy: having more than one spouse or partner at a time (males or females; adj: polygamous).

Polymorphism: a variation at a specific location on a genetic sequence.

Polyspermy: an egg fertilized by more than one sperm, usually lethal to that egg.

Precocial: born or hatched in a relatively developed state, such as to require limited parental care. *See* **Atrical**.

Pseudo-competition: the appearance that a species outcompetes its relatives for resources based on misdirected and aggressive courtship behaviours.

Pseudo-copulation: misdirected sexual behaviours, like those of certain insects which attempt to mate with female-mimicking flowers.

Pseudopenises: external genitalia that resemble penises, like the penile-clitoris in female hyenas.

Pseudoscrotum: a non-testes-containing body part that resembles a scrotum; e.g. the fused labia in female hyenas.

Quiescent: inactive or dormant.

Senescence: age-related deterioration. Can be actuarial (of the body) or reproductive (ability to produce offspring).

Setae: a stiff hair or bristle-like structure. Found in many invertebrates, but especially segmented worms (members of the phylum Annelida).

Sexual monogamy: the behaviour of copulating with and producing offspring from one and only one partner; may or may not be accompanied by social monogamy.

Siblicide: the killing of a sibling. *See* **Infanticide**.

Sire: to be the male parent (father).

Social monogamy: the behaviour of having one and only one partner for raising young; can be during a breeding season or for life, and may or may not be accompanied by sexual monogamy.

Sociosexual: relating to interpersonal or *social* aspects of sexuality.

Somatic: adjective for non-reproductive (e.g. somatic cells are all the body cells except for those that produce eggs and sperms).

Sperm: the mature male reproductive cell (gamete); spermatozoa. *See* **Gamete**.

Spermalege: a highly modified region of the abdomen of female bed bugs where the male usually punctures her during copulation.

Spermatheca: a kind of female sperm-storage organ in invertebrates.

Spermatophore: a sac which contains sperm, transferred from the male to the female during copulation.

Spermatophylax: a gelatinous material secreted around the spermatophore that the female consumes.

STI: sexually transmitted infection, which can refer to bacteria, viruses, or parasites that are passed from one individual to another during intercourse.

Stochastic: random, not at regular intervals, not characterized by a fixed rate or predictable pattern. Stochasticity: randomness; that which varies related to chance alone.

Sympatric: possessing ranges with significant geographic overlap.

Taxonomic: of or related to the classification of organisms (taxonomy).

Terminal investment: an increase in reproductive effort due to perceived or actual impending death; can be age-related or due to illness.

Thanatosis: a death-like state characterized by the cessation of movement and other voluntary activities, usually accompanied by the assuming a posture that suggests mortality.

Traumatic insemination: sperm delivery method where the male physically pierces the female to inject his ejaculate, resulting in minor or major bodily harm.

Vestigial: a remnant organ or body part which is either smaller, degenerate, or no longer functions as it once did.

Viviparity: reproduction through the birthing of live young (adj: viviparous). *See* **Oviparity**.

Zygote: a fertilized egg; an embryo which can now undergo development; the product of the union of a male gamete (sperm) and a female gamete (ovum or egg). *See* **Gamete**, **Ovum**, **Sperm**.

BIBLIOGRAPHY

SECTION ONE: THE MEET

Call me maybe

Holly L. Hennin, Nicole K. S. Barker, David W. Bradley, Daniel J. Mennill, 'Bachelor and paired male rufous-and-white wrens use different singing strategies', *Behavioral Ecology and Sociobiology*, December 2009, 64 (2): 151–9.

Christopher N. Templeton, Alejandro A. Ríos-Chelén, Esmeralda Quirós-Guerrero, Nigel I. Mann, Peter J. B. Slater, 'Female happy wrens select songs to cooperate with their mates rather than confront intruders', *Biology Letters*, 2013, 9: 20120863.

Michelle L. Hall, Anne Peters, 'Do male paternity guards ensure female fidelity in a duetting fairy-wren?', *Behavioral Ecology*, 2009, 20: 222–8.

Mitch A. Tucker, H. C. Gerhardt, 'Parallel changes in mate-attracting calls and female preferences in autotriploid tree frogs', *Proceedings of the Royal Society B; Biological Sciences*, 2012, 279: 1583–7.

B. Lardner, M. bin Lakim, 'Tree-hole frogs exploit resonance effects', *Nature*, 2002, Vol. 420: 475.

Machteld N. Verzijden, Jasper van Heusden, Niels Bouton, Frans Witte, Carel ten Cate, Hans Slabbekoorna, 'Sounds of male Lake

Victoria cichlids vary within and between species and affect female mate preferences', *Behavioral Ecology*, 2010, 21: 548–55.

Jianguo Cui, Yezhong Tang, Peter M. Narins, 'Real estate ads in Emei music frog vocalizations: female preference for calls emanating from burrows', *Biology Letters*, 2012, 8:337–40.

Nicole Geberzahn, Wolfgang Goymann, Carel ten Cate, 'Threat signaling in female song—evidence from playbacks in a sex-role reversed bird species', *Behavioral Ecology*, 2010, 21: 1147–55.

Nicole Geberzahn, Wolfgang Goymann, Christina Muck, Carel ten Cate, 'Females alter their song when challenged in a sex-role reversed bird species', *Behavioral Ecology*, 2009, 64:193–204

Samuel P. Caro, Kendra B. Sewall, Katrina G. Salvante, and Keith W. Sockman, 'Female Lincoln's sparrows modulate their behavior in response to variation in male song quality', *Behavioral Ecology*, 2010, 21: 562–9.

Björn Lardner, Maklarin B. Lakim, 'Female call preferences in tree-hole frogs: why are there so many unattractive males?', *Animal Behaviour*, 2004, 68: 265–72.

A little chemical help

Martinus E. Huigens, Jozef B. Woelke, Foteini G. Pashalidou, T. Bukovinszky, Hans M. Smid, Nina E. Fatouros, 'Chemical espionage on species-specific butterfly anti-aphrodisiacs by hitchhiking Trichogramma wasps', *Behavioral Ecology*, 2010, 21: 470–8.

Marianne Peso, Mark A. Elgar, Andrew B. Barron, 'Pheromonal control: reconciling physiological mechanism with signalling theory', *Biological Reviews*, 2014, 542–59.

Joachim Ruther & Theresa Hammerl, 'An oral male courtship pheromone terminates the response of Nasonia vitripennis females to the male-produced sex attractant', *Journal of Chemical Ecology*, 2014, 40: 56–62.

Michael R. Kidd, Peter D. Dijkstra, Callison Alcott, Dagan Lavee, Jacqualine Ma, Lauren A. O'Connell, Hans A. Hofmann,

'Prostaglandin F2α facilitates female mating behavior based on male performance', *Behavioral Ecology*, 2013, 67: 1307–15.

T. Pokorny, M. Hannibal, J. J. G. Quezada-Euan, E. Hedenström, N. Sjöberg, J. Bang, T. Eltz, 'Acquisition of species-specific perfume blends: influence of habitat-dependent compound availability on odour choices of male orchid bees (*Euglossa* spp.)', *Oecologia*, 2013, 172: 417–25.

R. Shine, T. Langkilde, R. T. Mason, 'Facultative pheromonal mimicry in snakes: "she-males" attract courtship only when it is useful', *Behavioral Ecology*, 2012, 66: 691–5.

Pedro Ere, Disconzi Brum, Luiz Ernesto Costa-Schmidt, Aldo Mellender de Araujoa, 'It is a matter of taste: chemical signals mediate nuptial gift acceptance in a neotropical spider', *Behavioral Ecology*, 2012, 23: 442–7.

Melissa L. Thomas, Leigh W. Simmons, 'Male dominance influences pheromone expression, ejaculate quality, and fertilization success in the Australian field cricket, Teleogryllus oceanicus', *Behavioral Ecology*, 2009, 20: 1118–24.

Glenn P. Svensson, Eylem Akman Gündüz, Natalia Sjöberg, Erik Hedenström, Jean-Marc Lassance, Hong-Lei Wang, Christer Löfstedt, Olle Anderbrant, 'Identification, synthesis, and behavioral activity of 5,11-Dimethylpentacosane, a novel sex pheromone component of the greater wax moth, Galleria Mellonella (L.)', *Journal of Chemical Ecology*, 2014, 40: 387–95.

Simon Vitecek, Annick Maria, Catherine Blais, Line Duportets, Cyril Gaertner, Marie-Cécile Dufour, David Siaussat, Stéphane Debernard, Christophe Gadenne, 'Is the rapid post-mating inhibition of pheromone response triggered by ecdysteroids or other factors from the sex accessory glands in the male moth Agrotis ipsilon?', *Hormones and Behavior*, 2013, 63: 700–708.

Mirjam Amcoff, Lara R. Hallsson, Svante Winberg, Niclas Kolm, 'Male courtship pheromones affect female behaviour in the swordtail Characin (Corynopoma riisei)', *Ethology*, 2014, 120: 463–70.

Stefan H. Nessler, Gabriele Uhl, Jutta M. Schneider, 'Scent of a woman – the effect of female presence on sexual cannibalism in an orb-weaving spider (Araneae: Araneidae)', *Ethology*, 2009, 115: 633–40.

Fernando da Silva Carvalho Filho, 'Scent-robbing and fighting among male orchid bees, *Eulaema* (Apeulaema) *nigrita* Lepeletier, 1841 (Hymenoptera: Apidae: Euglossini)', *Biota Neotrop*, 10(2).

Y. Zimmermann, D. W. Roubik, J. J. G. Quezada-Euan, R. J. Paxton, T. Eltz, 'Single mating in orchid bees (Euglossa, Apinae): implications for mate choice and social evolution', *Insect Society*, 2009, 56: 241–9.

Luciana Baruffaldi, Fernando G. Costa, 'Male reproductive decision is constrained by sex pheromones produced by females', *Behaviour*, 2014, 151: 465–77.

What's your sign?

Simona Kralj-Fišer, Graciela A. Sanguino Mostajo, Onno Preik, Stano Pekár, Jutta M. Schneider, 'Assortative mating by aggressiveness type in orb weaving spiders', *Behavioral Ecology*, 2013, 10: 1093.

Kasey D. Fowler-Finn, Emilia Triana, Owen G. Miller, 'Mating in the harvestman *Leiobunum vittatum* (Arachnida: Opilionidae): from premating struggles to solicitous tactile engagement', *Behaviour*, 2014, DOI:10.1163/1568539X-00003209

Afiwa Midamegbe, Arnaud Gregoire, Vincent Staszewski, Philippe Perret, Marcel M. Lambrechts, Thierry Boulinier, Claire Doutrelant, 'Female blue tits with brighter yellow chests transfer more carotenoids to their eggs after an immune challenge', *Oecologia*, 2013, 173: 387–97.

Jen Crick, Malini Suchak, Timothy M. Eppley, Matthew W. Campbell, Frans B.M. de Waal, 'The roles of food quality and sex in chimpanzee sharing behavior (Pan *troglodytes)*', *Behaviour*, 2013, 150: 1203–24.

Joseph L. Tomkins, Wade N. Hazel, Marissa A. Penrose, Jacek W. Radwan, and Natasha R. LeBas, 'Habitat complexity drives experimental evolution of a conditionally expressed secondary sexual trait', *Current Biology*, 2011, 21: 569–73.

Jonathan N. Pruitt, Susan E. Riechert, 'How within-group behavioural variation and task efficiency enhance fitness in a social group', *Proceedings of the Royal Society B: Biological Sciences*, 2011, 278: 1209–15.

Samuel D. Gosling, Oliver P. John, 'Personality dimensions in nonhuman animals: a cross-species review', *Current Directions in Psychological Science*, 1999, 8:69.

Sasha R. X. Dall, Alasdair I. Houston, John M. McNamara, 'The behavioural ecology of personality: consistent individual differences from an adaptive perspective', *Ecology Letters*, 2004, 7: 734–39.

Teresa L. Dzieweczynski, Alyssa M. Russell, Lindsay M. Forrette, Krystal L. Mannion, 'Male behavioral type affects female preference in Siamese fighting fish', *Behavioral Ecology*, 2014, 25(1): 136–41.

Jason V. Watters, 'Can the alternative male tactics "fighter" and "sneaker" be considered "coercer" and "cooperator" in coho salmon?', *Animal Behaviour*, 2005, 70: 1055–62.

Alexander G. Ophir, Bennett G. Galef, Jr, 'Female Japanese quail that "eavesdrop" on fighting males prefer losers to winners', *Animal Behaviour*, 2003, 66: 399–407.

Wiebke Schuett, Jean-Guy J. Godin, Sasha R. X. Dall, 'Do female zebra finches, taeniopygia guttata, choose their mates based on their "personality"?', *Ethology*, 2011, 117: 908–17.

Wiebke Schuett, Jesse Laaksonen, Toni Laaksonen, 'Prospecting at conspecific nests and exploration in a novel environment are associated with reproductive success in the jackdaw', *Behavioral Ecology and Sociobiology*, September 2012, 66(9): 1341–50.

Wiebke Schuett, Tom Tregenza, Sasha R. X. Dall, 'Sexual selection and animal personality', *Biological Reviews*, 2010, 85: 217–46.

Jonathan N. Pruitt, Susan E. Riechert, David J. Harris, 'Reproductive consequences of male body mass and aggressiveness depend on females' behavioral types', *Behavioral Ecology and Sociobiology*, 2011, 65: 1957–66.

Plastic partners

Nils Cordes, Leif Engqvist, Tim Schmoll, Klaus Reinhold, 'Sexual signaling under predation: attractive moths take the greater risks', *Behavioral Ecology*, 2014, 25(2): 409–14.

Christina Richardson, Thierry Lengagne, 'Multiple signals and male spacing affect female preference at cocktail parties in treefrogs', *Proceedings of the Royal Society B: Biological Sciences*, 2010, 277: 1247–52.

Karla M. Addessoa, Katherine A. Short, Allen J. Moore, Christine W. Miller, 'Context-dependent female mate preferences in leaf-footed cactus bugs', *Behaviour*, 2014, 151: 479–92.

Luke Holman, Andrew T. Kahn, Patricia R.Y. Backwell, 'Fiddlers on the roof: elevation muddles mate choice in fiddler crabs', *Behavioral Ecology*, 2014, 25(2): 271–5.

Karin Gross, Gilberto Pasinelli, Hansjoerg P. Kune, 'Behavioral plasticity allows short-term adjustment to a novel environment', *The American Naturalist*, 2010, 176: 456–64.

Ulrike Lampe, Tim Schmoll, Alexandra Franzke, Klaus Reinhold, 'Staying tuned: grasshoppers from noisy roadside habitats produce courtship signals with elevated frequency components', *Functional Ecology*, 2012, 26: 1348–54.

Paula Sicsú. Lilian T. Manica, Rafael Maia, Regina H. Macedo, 'Here comes the sun: multimodal displays are associated with sunlight incidence', *Behavioral Ecology and Sociobiology*, 2013, 67: 1633–42.

F. Stephen Dobson, 'Live fast, die young, and win the sperm competition', *PNAS*, 29 October 2013, 110 (44): 17610–11.

Terry J. Ord, Richard A. Peters, Barbara Clucas, Judy A. Stamps, 'Lizards speed up visual displays in noisy motion habitats', *Proceedings of the Royal Society B: Biological Sciences*, 2007, 274: 1057–62.

Anna M. Billing, Gunilla Rosenqvist, Anders Berglund, 'No terminal investment in pipefish males: only young males exhibit risk-prone courtship behavior', *Behavioral Ecology*, 2007, 18: 535–40.

Clinton D. Francis, Catherine P. Ortega, Alexander Cruz, 'Different behavioural responses to anthropogenic noise by two closely related passerine birds', *Biology Letters*, 2011, 7: 850–52.

Natalie Pilakouta, Suzanne H. Alonzo, 'Predator exposure leads to a short-term reversal in female mate preferences in the green swordtail, *Xiphophorus helleri*', *Behavioral Ecology*, 2014, 25(2), 306–12.

Darrell J. Kemp, 'Sexual selection constrained by life history in a butterfly', *Proceedings of the Royal Society Lond. B*, 2002, 269: 1341–45.

Ulrika Candolin, 'Reproduction under predation risk and the trade-off between current and future reproduction in the threespine stickleback', *Proceedings of the Royal Society Lond. B*, 1998, 265: 1171–5.

E. W. Ruell, C. A. Handelsman, C. L. Hawkins, H. R. Sofaer, C. K. Ghalambor, L. Angeloni, 'Fear, food and sexual ornamentation: plasticity of colour development in Trinidadian guppies', *Proceedings of the Royal Society Lond. B*, 1998, 280: 20122019.

Kasey D. Fowler-Finn, Eileen A. Hebets, 'The degree of response to increased predation risk corresponds to male secondary sexual traits', *Behavioral Ecology*, 2011, 22: 268–75.

Jan Heuschelea, Tiina Salminena, Ulrika Candolina, 'Habitat change influences mate search behaviour in three-spined sticklebacks', *Animal Behaviour*, 2012, 83: 1505–10.

Lisa A. Taylor, Kevin J. McGraw, 'Male ornamental coloration improves courtship success in a jumping spider, but only in the sun', *Behavioral Ecology*, 2013, 24(4): 955–67.

Robert J. P. Heathcote, Emily Bell, Patrizia d'Ettorre, Geoffrey M. While, Tobias Uller, 'The scent of sun worship: basking experience alters scent mark composition in male lizards', *Behavioral Ecology and Sociobiology*, 2014, 68: 861–70.

Gonçalo C. Cardoso, Jonathan W. Atwell, 'On the relation between loudness and the increased song frequency of urban birds', *Animal Behaviour*, 2011, 82: 831–6.

SuperTramps

Malte Andersson, Matti Ahlund, 'Don't put all your eggs in one nest: spread them and cut time at risk', *The American Naturalist*, 2012, 180 (3): 354–63.

Elodie F. Briefer, Mary E. Farrell, Thomas J. Hayden, Alan G. McElligott, 'Fallow deer polyandry is related to fertilization insurance', *Behavioral Ecology and Sociobiology*, 2013, 67: 657–65.

Panu Välimäki, Sami M. Kivelä, Maarit I. Mäenpää, 'Mating with a kin decreases female remating interval: a possible example of inbreeding avoidance', *Behavioral Ecology and Sociobiology*, 2011, 65: 2037–47.

Demian D. Chapman, Sabine P. Wintner, Debra L. Abercrombie, Jimiane Ashe, Andrea M. Bernard, Mahmood S. Shivji, Kevin A. Feldheim, 'The behavioural and genetic mating system of the sand tiger shark, *Carcharias taurus*, an intrauterine cannibal', *Biology Letters*, 2013, 9: 20130003

Stephanie J. Kamel, Richard K. Grosberg, 'Exclusive male care despite extreme female promiscuity and low paternity in a marine snail,' *Ecology Letters*, 2012, 15: 1167–73.

Elise Huchard, Cindy I. Canale, Chloe Le Gros, Martine Perret, Pierre-Yves Henry, and Peter M. Kappeler, 'Convenience polyandry or convenience polygyny? Costly sex under female control in a promiscuous primate', *Proceedings of the Royal Society B*, 2012, 279: 1371–79.

Marcel P. Haesler, Charlotte M. Lindeyer, Oliver Otti, Danielle Bonfils, Dik Heg, and Michael Taborskya, 'Female mouthbrooders in control of pre- and postmating sexual selection', *Behavioral Ecology*, 2011, 22: 1033–41.

Barbara A. Caspers, E. Tobias Krause, Ralf Hendrix, Michael Kopp, Oliver Rupp, Katrin Rosentreter, Sebastian Steinfartz, 'The more the better – polyandry and genetic similarity are positively linked to reproductive success in a natural population of terrestrial salamanders (*Salamandra salamandra*)', *Molecular Ecology*, 2014, 23: 239–50.

Diana O. Fisher, Michael C. Double, Simon P. Blomberg, Michael D. Jennions, Andrew Cockburn, 'Post-mating sexual selection increases lifetime fitness of polyandrous females in the wild', *Nature*, 2006, 444: 89–92.

Jerry O. Wolff, David W. Macdonald, 'Promiscuous females protect their offspring', *Trends in Ecology and Evolution*, 2004, 19(3): 127–34.

Lucy I. Wright, Wayne J. Fuller, Brendan J. Godley, Andrew McGowan, Tom Tregenza, Annette C. Broderick, 'No benefits of polyandry to female green turtles', *Behavioral Ecology*, 2013, 24: 1022–9.

Monogamy. Really?

M. Genevieve W. Jones, N. M. S. Mareile Techow, Peter G. Ryan, 'Dalliances and doubtful dads: what determines extra-pair paternity in socially monogamous wandering albatrosses?', *Behavioral Ecology and Sociobiology*, 2012, 66: 1213–24.

Miyako H. Warrington, Lee Ann Rollins, Nichola J. Raihani, Andrew F. Russell, Simon C. Griffith, 'Genetic monogamy despite variable ecological conditions and social environment in the cooperatively breeding apostlebird', *Ecology and Evolution*, 2013, 3(14): 4669–82.

Cassandra Cameron, Dominique Berteaux, France Dufresne, 'Spatial variation in food availability predicts extrapair paternity in the arctic fox', *Behavioral Ecology*, 2011, 22: 1364–73.

Dieter Lukas, Tim Clutton-Brock, 'Evolution of social monogamy in primates is not consistently associated with male infanticide', *PNAS*, 29 April 2014, 111(17): E1674.

Christopher Opie, Quentin D. Atkinson, Robin I. M. Dunbar, Susanne Shultz, 'Reply to Lukas and Clutton-Brock: Infanticide still drives primate monogamy', *PNAS*, 29 April 2014, 111(17): E1675.

Xavier A. Harrison, Jennifer E. York, Dominic L. Cram, Michelle C. Hares, Andrew J. Young, 'Complete reproductive skew within white-browed sparrow weaver groups despite outbreeding opportunities for subordinates of both sexes', *Behavioral Ecology and Sociobiology*, 2013, 67: 1915–29.

Maria R. Servedio, Trevor D. Price, Russell Lande, 'Evolution of displays within the pair bond', *Proceedings of the Royal Society B*, 2013, 280: 20123020.

Irene M. van den Heuvel, Michael I. Cherry, Georg M. Klump, 'Crimson-breasted Shrike females with extra pair offspring contributed more to duets', *Behavioral Ecology and Sociobiology*, 2014, 68: 1245–52.

Katrine S. Hoset, Yngve Espmark, Frode Fossøy, Bård G. Stokke, Henrik Jensen, Morten I. Wedege, Arne Moksnes, 'Extra-pair paternity in relation to regional and local climate in an Arctic-breeding passerine', *Polar Biology*, 2014, 37: 89–97.

Aneta Arct, Szymon M. Drobniak, Edyta Podmokła, Lars Gustafson, Mariusz Cichoń, 'Benefits of extra-pair mating may depend on environmental conditions—an experimental study in the blue tit (*Cyanistes caeruleus*)', *Behavioral Ecology and Sociobiology*, 2013, 67: 1809–15.

Andrew M. Robbins, Maryke Gray, Augustin Basabose, Prosper Uwingeli, Innocent Mburanumwe, Edwin Kagoda, Martha M. Robbins, 'Impact of male infanticide on the social structure of mountain gorillas', *PLOS ONE*, November 2013, 8 (11): e78256.

Lance G. Woolaver, Rina K. Nichols, Eugene S. Morton, Bridget J. M. Stutchbury, 'Social and genetic mating system of Ridgway's hawk (Buteo *ridgwayi*), an endemic raptor on Hispaniola', *Journal of Tropical Ecology*, November 2013, 29 (06): 531–40.

Marie J. E. Charpentier, Christine M. Drea, 'Victims of infanticide and conspecific bite wounding in a female-dominant primate: a long-term study', *PLOS ONE*, December 2013, 8 (12): e82830.

Alan F. Dixson, 'Male infanticide and primate monogamy', *PNAS*, 17 December 2013, 110 (51), E4937.

Andras Liker, Robert P. Freckleton, Tamas Szekely, 'Divorce and infidelity are associated with skewed adult sex ratios in birds', *Current Biology*, 14 April 2014, 24: 880–84.

E. H. DuVal, 'Female mate fidelity in a lek mating system and its implications for the evolution of cooperative lekking behavior', *The American Naturalist*, February 2013, 181 (2): 213–222.

Oscar Sánchez-Macouzet, Cristina Rodríguez, Hugh Drummond, 'Better stay together: pair bond duration increases individual fitness independent of age-related variation', *Proceedings of the Royal Society B*, 2014, 281: 201232843.

Dieter Lukas, Tim Clutton-Brock, 'Cooperative breeding and monogamy in mammalian societies', *Proceedings of the Royal Society B*, 2012, 279: 20112468.

Christopher Opie, Quentin D. Atkinson, Robin I. M. Dunbar, Susanne Shultz, 'Male infanticide leads to social monogamy in primates', *PNAS*, 13 August 2013, 110 (33): 13328–32.

Charlie K. Cornwallis, Stuart A. West, Katie E. Davis, Ashleigh S. Griffin, 'Promiscuity and the evolutionary transition to complex societies', *Nature*, 19 August 2010, 466: 969.

William O. H. Hughes *et al*, 'Ancestral monogamy shows kin selection is key to the evolution of eusociality', *Science*, 2008, 320: 1213.

Frans B. M. de Waal, Sergey Gavrilets, 'Monogamy with a purpose', *PNAS*, 17 September 2013, 110 (38), 15167–8.

Maren Huck, Eduardo Fernandez-Duque, Paul Babb, Theodore Schurr, 'Correlates of genetic monogamy in socially monogamous mammals: insights from Azara's owl monkeys', *Proceedings of the Royal Society B*, 2014, 281: 20140195.

Andrea K. Townsend, Reed Bowman, John W. Fitzpatrick, Michelle Dent, Irby J. Lovette, 'Genetic monogamy across variable demographic landscapes in cooperatively breeding Florida scrub-jays', *Behavioral Ecology*, 2011, 22: 464–70.

J. Antonio Baeza, Juan A. Bolaños, Jesús E. Hernandez, Carlos Lira, Régulo López, 'Monogamy does not last long in Pontonia mexicana, a symbiotic shrimp of the amber pen-shell Pinna carnea from the southeastern Caribbean Sea', *Journal of Experimental Marine Biology and Ecology*, 2011, 407: 41–47.

Peter Nonacs, 'Resolving the evolution of sterile worker castes: a window on the advantages and disadvantages of monogamy', *Biology Letters*, 2014, 10: 20140089.

Pia O. Gabriel, Jeffrey M. Black, 'Correlates and Consequences of the Pair Bond in Steller's Jays', *Ethology*, 2013, 119: 178–187.

Andréanne Lessard, Audrey Bourret, Marc Bélisle, Fanie Pelletier, Dany Garant, 'Individual and environmental determinants of reproductive success in male tree swallow (*Tachycineta bicolor*)', *Behavioral Ecology and Sociobiology*, 2014, 68: 733–42.

Trinkets and tokens

Susan N. Gershman, Christopher Mitchell, Scott K. Sakaluk, John Hunt, 'Biting off more than you can chew: sexual selection on the free amino acid composition of the spermatophylax in decorated crickets', *Proceedings of the Royal Society B*, 2012, 279: 2531–8.

Bonnie J. Holmes, David T. Neil, '"Gift giving" by wild bottle-nose dolphins (*Tursiops sp.*) to humans at a wild dolphin provisioning program, Tangalooma, Australia', *Anthrozoos*, 2012, 25(4): 397–413.

'Christian Luis Rodriguez-Enriquez, Eduardo Tadeo, Juan Rull, 'Elucidating the function of ejaculate expulsion and consumption after copulation by female *Euxesta bilimeki*', *Behavioral Ecology and Sociobiology*, 2013, 67: 937–46.

S. N. Gershman, J. Hunt, S. K. Sakaluk, 'Food fight: sexual conflict over free amino acids in the nuptial gifts of male decorated crickets', *Journal of Evolutionary Biology*, 2013, 26: 693–704.

Maria J. Albo, Trine Bilde, Gabriele Uhl, 'Sperm storage mediated by cryptic female choice for nuptial gifts', *Proceedings of the Royal Society B*, 2013, 280: 20131735.

Natasha Tigreros, Emma M. Sass, Sara M. Lewis, 'Sex-specific response to nutrient limitation and its effects on female mating success in a gift-giving butterfly', *Evolutionary Ecology*, 2013, 27: 1145–58.

Karim Vahed, 'All that Glisters is not Gold: Sensory Bias, Sexual Conflict and Nuptial Feeding in Insects and Spiders', *Ethology*, 2007, 113: 105–127.

C. Tuni, M. J. Albo, T. Bilde, 'Polyandrous females acquire indirect benefits in a nuptial feeding species', *Journal of Evolutionary Biology*, 2013, 26: 1307–16.

Darryl T. Gwynne, 'Sexual conflict over nuptial gifts in insects', *Annual Review of Entomology*, 2008, 53: 83–101.

Pavol Prokop, Michael R. Maxwell, 'Gift carrying in the spider *Pisaura mirabilis*: nuptial gift contents in nature and effects on male running speed and fighting success', *Animal Behaviour*, 2012, 83: 1395–99.

Paolo Giovanni Ghislandi, Maria J. Albo, Cristina Tuni, Trine Bilde, 'Evolution of deceit by worthless donations in a nuptial gift-giving spider', *Current Zoology*, 2014, 60 (1): 43–51.

I'll have what she's having

Yukio Matsumotoa, Takeshi Takegaki, 'Female mate choice copying increases egg survival rate but does not reduce mate-sampling cost in the barred-chin blenny', *Animal Behaviour*, 2013, 86: 339–46.

Laurence Henry, Cécile Bourguet, Marion Coulon, Christine Aubry, Martine Hausberger, 'Sharing mates and nest boxes is associated with female "friendship" in European Starlings, *Sturnus vulgaris*', *Journal of Comparative Psychology*, 2013, 127 (1): 1–13.

Gerlinde Höbel, Robb C. Kolodziej, 'Wood frogs (Lithobates *sylvaticus)* use water surface waves in their reproductive behaviour', *Behaviour*, 2013, 150: 471–83.

M. M. Webster, K. N. Laland, 'Local enhancement via eavesdropping on courtship displays in male guppies, *Poecilia reticulata*', *Animal Behaviour*, 2013, 86: 75–83.

Torsten Wronski, David Bierbach, Lara-Marlene Czupalla, Hannes Lerp, Madlen Ziege, Peter L. Cunningham, Martin Plath, 'Rival presence leads to reversible changes in male mate choice of a desert dwelling ungulate', *Behavioral Ecology*, 2012, 23: 551–8.

David Bierbach, Christian T. Jung, Simon Hornung, Bruno Streit, Martin Plath, 'Homosexual behaviour increases male attractiveness to females', *Biology Letters*, 2013, 9: 20121038.

Rachel L. Moran, Carl N. von Ende, Bethia H. King, 'Mate choice copying in two species of darters (Percidae: Etheostoma)', *Behaviour*, 2013, 150: 1255–74.

D. Waynforth, 'Mate choice copying in humans', *Human Nature*, 2007, 18: 264–71.

Robert I. Bowers, Skyler S. Place, Peter M. Todd, Lars Penke, Jens B. Asendorpf, 'Generalization in mate-choice copying in humans', *Behavioral Ecology*, 2012, 23: 112–24.

Jessica Parker, Melissa Burkley, 'Who's chasing whom? The impact of gender and relationship status on mate poaching', *Journal of Experimental Social Psychology*, 2009, 45: 1016–19.

Roslyn Dakin, Robert Montgomerie, 'Deceptive copulation calls attract female visitors to peacock leks', *The American Naturalist*, 2014, 183 (4): 558–64.

Nice guys finish last

Scott L. Applebaum, Alexander Cruz, 'The role of mate-choice copying and disruption effects in mate preference determination of *Limia perugiae* (Cyprinodontiformes, Poeciliidae)', *Ethology*, 2000, 106: 933–44.

Marina Magaña, Juan C. Alonso, Carlos Palacín, 'Age-related dominance helps reduce male aggressiveness in great bustard leks', *Animal Behaviour*, 2011, 82: 203–11.

Stein Are Saether, Peder Fiske, John Atle Kalas, 'Pushy males and choosy females: courtship disruption and mate choice in the lekking great snipe', *Proceedings of the Royal Society London B*, 1999, 266: 1227–34.

David B. Lank, Constance M. Smith, Olivier Hanotte, Arvo Ohtonen, Simon Bailey, Terry Burke, 'High frequency of polyandry in a lek mating system', *Behavioral Ecology*, 2002, 13: 209–15.

Andrew Balmford, 'Mate choice on leks', *Tree*, March 1991, 6 (3): 87–92.

Stephen Pruett-Jones, Melinda Pruett-Jones, 'Sexual competition and courtship disruptions: why do male bowerbirds destroy each other's bowers?', *Animal Behaviour*, 1994, 47: 607–20.

Michael S. Webster, Scott K. Robinson, 'Courtship disruptions and male mating strategies: examples from female-defense mating systems', *The American Naturalist*, December 1999, 154 (6): 717–29.

Mercedes S. Foster, 'Disruption, dispersion and dominance in lek-breeding birds', *The American Naturalist*, July 1983, 122 (1): 53–72.

Pepper W. Trail, 'Courtship disruption modifies mate choice in a lek-breeding bird', *Science*, 15 February 1985, 227 (4688): 778–80.

Joah R. Madden, Tamsin J. Lowe, Hannah V. Fuller, Rebecca L. Coe, Kanchon K. Dasmahapatra, William Amos, Francine Jury, 'Neighbouring male spotted bowerbirds are not related, but do maraud each other', *Animal Behaviour*, 2004, 68: 751–8.

Stephen Pruett-Jones, Aviad Heifetz, 'Optimal marauding in bowerbirds', *Behavioral Ecology*, 2012, 23: 607–14.

Sin-Yeon Kim, Alberto Velando, 'Stickleback males increase red coloration and courtship behaviours in the presence of a competitive rival', *Ethology*, 2014, 120: 502–10.

Fool me once

Laura A. Kelley, John A. Endler, 'Male great bowerbirds create forced perspective illusions with consistently different individual quality', *PNAS*, 18 December 2012, 109 (51): 20980–20985.

N. J. Vereecken, J. N. McNeil, 'Cheaters and liars: chemical mimicry at its finest', *Canadian Journal of Zoology*, 2010, 88: 725–52.

Christina Richardson, Thierry Lengagne, 'Multiple signals and male spacing affect female preference at cocktail parties in treefrogs', *Proceedings of the Royal Society B*, 2010, 277: 1247–52.

Roslyn Dakin, Robert Montgomerie, 'Deceptive copulation calls attract female visitors to peacock leks', *The American Naturalist*, 2014, 183 (4): 558–64.

Jakob Bro-Jørgensen, Wiline M. Pangle, 'Male topi antelopes alarm snort deceptively to retain females for mating', *The American Naturalist*, 2010, 176 (1): E33–9.

Candice L. Bywater, Robbie S. Wilson, 'Is honesty the best policy? Testing signal reliability in fiddler crabs when receiver-dependent costs are high', *Functional Ecology*, 2012, 20: 804–11.

Aliza le Roux, Noah Snyder-Mackler, Eila K. Roberts, Jacinta C. Beehner, Thore J. Bergman, 'Evidence for tactical concealment in a wild primate', *Nature Communications*, 12 February 2013

Christopher H. Martin, 'Unexploited females and unreliable signals of male quality in a Malawi cichlid bower polymorphism', *Behavioral Ecology*, 2010, 21: 1195–1202.

William E Wagner, Jr, Andrew R Smith, Alexandra L Basolo, 'False promises: females spurn cheating males in a field cricket', *Biology Letters*, 2007, 3: 379–81.

Marinus L. de Jager, Allan G. Ellis, 'Costs of deception and learned resistance in deceptive interactions', *Proceedings of the Royal Society B*, 2014, 281: 20132861.

James E. Lloyd, 'Male photuris fireflies mimic sexual signals of their females' prey', *Science*, 7 November 1980, 210 (4470): 669–71.

David C. Marshall, Kathy B. R. Hill, 'Versatile Aggressive Mimicry of Cicadas by an Australian Predatory Katydid', *PLOS ONE*, January 2009, 4 (1): e4185.

Leslie S. Saul-Gershenz, Jocelyn G. Millar, 'Phoretic nest parasites use sexual deception to obtain transport to their host's nest', *PNAS*, 19 September 2006, 103 (38): 14039–44.

Ryan D. Phillips, Tingbao Xu, Michael F. Hutchinson, Kingsley W. Dixon, Rod Peakall, 'Convergent specialization – the sharing of pollinators by sympatric genera of sexually deceptive orchids', *Journal of Ecology*, 2013, 101: 826–35.

Florian P. Schiestl, Manfred Ayasse, Hannes F. Paulus, Christer

Löfstedt, Bill S. Hansson, Fernando Ibarra, Wittko Francke, 'Orchid pollination by sexual swindle', *Nature*, 3 June 1999, 399: 421–2

Kyle Summers, 'Sexual conflict and deception in poison frogs', *Current Zoology*, 2014, 60 (1): 37–42.

Jussi Lehtonen, Michael R. Whitehead, 'Sexual deception: coevolution or inescapable exploitation?', *Current Zoology*, 2014, 60 (1): 52–61.

Secrets and lies

N. J. Vereecken, J. N. McNeil, 'Cheaters and liars: chemical mimicry at its finest', *Canadian Journal of Zoology*, 2010, 88: 725–52.

Marinus L. de Jager, Allan G. Ellis, 'Costs of deception and learned resistance in deceptive interactions', *Proceedings of the Royal Society B*, 2014, 281: 20132861.

Nicolas J. Vereecken, Carol A. Wilson, Susann Hötling, Stefan Schulz, Sergey A. Banketov, Patrick Mardulyn, 'Pre-adaptations and the evolution of pollination by sexual deception: Cope's rule of specialization revisited', *Proceedings of the Royal Society B*, 2012, 279: 4786–94.

A. C. Gaskett, 'Orchid pollination by sexual deception: pollinator perspectives', *Biological Reviews*, 2011, 86: 33–75.

Agustín Sanguinetti, Rodrigo Bustos Singer, 'Invasive bees promote high reproductive success in Andean orchids', *Biological Conservation*, 2014, 175: 10–20.

Ryan D. Phillips, Tingbao Xu, Michael F. Hutchinson, Kingsley W. Dixon, Rod Peakall, 'Convergent specialization – the sharing of pollinators by sympatric genera of sexually deceptive orchids', *Journal of Ecology*, 2013, 101: 826–35.

Demetra Rakosy, Martin Streinzer, Hannes F. Paulus, Johannes Spaethe, 'Floral visual signal increases reproductive success in a sexually deceptive orchid', *Arthropod-Plant Interactions*, 2012, 6: 671–81.

A. C. Gaskett, C. G. Winnick, M. E. Herberstein, 'Orchid sexual deceit provokes ejaculation', *The American Naturalist*, June 2008, 171 (6): E206–12.

Jussi Lehtonen, Michael R. Whitehead, 'Sexual deception: coevolution or inescapable exploitation?', *Current Zoology*, 2014, 60 (1): 52–61.

Rod Peakall, Michael R. Whitehead, 'Floral odour chemistry defines species boundaries and underpins strong reproductive isolation in sexually deceptive orchids', *Annals of Botany*, 2014, 113: 341–55.

Ryan D. Phillips, Daniela Scaccabarozzi, Bryony A. Retter, Christine Hayes1, Graham R. Brown, Kingsley W. Dixon and Rod Peakall, 'Caught in the act: pollination of sexually deceptive trap-flowers by fungus gnats in *Pterostylis* (Orchidaceae), *Annals of Botany*, 2014, 113: 629–41.

M. M. Kelly, A. C. Gaskett, 'UV reflectance but no evidence for colour mimicry in a putative brood-deceptive orchid *Corybas cheesemanii*', *Current Zoology*, 2014, 60 (1): 104–13.

Sex or sick?

Alexandre Lerch, Lauriane Rat-Fischer, Laurent Nagle, 'Condition-dependent choosiness for highly attractive songs in female canaries', *Ethology*, 2012, 119: 58–65.

Emily K. Copeland, Kenneth M. Fedorka, 'The influence of male age and simulated pathogenic infection on producing a dishonest sexual signal', *Proceedings of the Royal Society B*, 2012, 279: 4740–46.

Sophie Beltran-Bech, Freddie-Jeanne Richard, 'Impact of infection on mate choice', *Animal Behaviour*, 2014, 90: 159–70.

Marie-Jeanne Holveck, Katharina Riebel, 'Low-quality females prefer low-quality males when choosing a mate', *Proceedings of the Royal Society B*, 2010, 277: 153–60.

Gabriel E. Leventhal, Robert P. Dünner, Seth M. Barribeau, 'Delayed virulence and limited costs promote fecundity compensation upon infection', *The American Naturalist*, April 2014, 183 (4): 480–93.

Deborah Pardo, Christophe Barbraud, Henri Weimerskirch, 'What shall I do now? State-dependent variations of life-history traits with aging in Wandering Albatrosses', *Ecology and Evolution*, 2014, 4 (4): 474–87.

Matthew W. H. Chatfield, Laura A Brannelly, Matthew J. Robak, Layla Freeborn,1 Simon P. Lailvaux, Corinne L. Richards-Zawacki, 'Fitness Consequences of Infection by *Batrachochytrium dendrobatidis* in Northern Leopard Frogs (*Lithobates pipiens*)', *EcoHealth*, 2013, 10: 90–8.

Jean M. Drayton, J. E. Kobus Boeke, Michael D. Jennions, 'Immune Challenge and Pre- and Post-copulatory Female Choice in the Cricket *Teleogryllus commodus*', *Journal of Inspect Behaviour*, 2013, 26: 176–90.

Pilar Lopez, Marianne Gabirot, Jose Martin, 'Immune challenge affects sexual coloration of male Iberian wall lizards', *Journal of Experimental Zoology*, 2009, 311A: 96–104.

Leigh W. Simmons, 'Resource allocation trade-off between sperm quality and immunity in the field cricket, *Teleogryllus oceanicus*', *Behavioral Ecology*, 2012, 23: 168–73.

S. S. French, G. I. H. Johnston, 'Immune activity suppresses reproduction in food-limited female tree lizards *Urosaurus ornatus*', *Functional Ecology*, 2007, 21: 1115–22.

Theo C. M. Bakker, Reto Künzler, Dominique Mazzi, 'Condition-related mate choice in sticklebacks', *Nature*, 16 September 1999, 401: 234.

Marie-Jeanne Holveck, Nicole Geberzahn, Katharina Riebel, 'An Experimental Test of Condition-Dependent Male and Female Mate Choice in Zebra Finches', *PLOS ONE*, August 2011, 6 (8): e23974.

SECTION TWO: THE SEX

Ejaculate!

Aldo Poiani, 'Complexity of seminal fluid: a review', *Behavioral Ecology and Sociobiology*, 2006, 60: 289–310.

Yumi Nakadera, Elferra M. Swart, Jeroen N.A. Hoffer, Onno den Boon, Jacintha Ellers, Joris M. Koene, 'Receipt of Seminal Fluid

Proteins Causes Reduction of Male Investment in a Simultaneous Hermaphrodite', *Current Biology*, 2014, 24: 859–62.

Lukas Schärer, 'Evolution: Don't be so butch, dear!', *Current Biology*, 2014, 24 (8): R311–13.

H. M. Tennant, E. E. Sonser, T. A. F. Long, 'Variation in male effects on female fecundity in *Drosophila melanogaster*', *Journal of Evolutionary Biology*, 2014, 27: 449–54.

Oliver Otti, Aimee P. McTighe, Klaus Reinhardt, '*In vitro* antimicrobial sperm protection by an ejaculate-like substance', *Functional Ecology*, 2013, 27: 219–26.

Klaus Reinhardt, Anne-Cecile Ribou, 'Females become infertile as the stored sperm's oxygen radicals increase', *Scientific Reports*, 2013, 3: 2888.

Jin Xu, Qiao Wang, 'Seminal fluid reduces female longevity and stimulates egg production and sperm trigger oviposition in a moth', *Journal of Insect Physiology*, 2011, 57: 385–90.

Jennifer C. Perry, Crystal T. Tse, 'Extreme costs of mating for male two-spot ladybird beetles', *PLOS ONE*, December 2013, 8 (12): e81934.

Jennifer C. Perry, Laura Sirot, Stuart Wigby, 'The seminal symphony: how to compose an ejaculate', *Trends in Ecology & Evolution*, July 2013, 28 (7): 414–22.

Luke Holman, '*Drosophila melanogaster* seminal fluid can protect the sperm of other males', *Functional Ecology*, 2009, 23: 180–6.

M. Aluja, J. Rull, J. Sivinski, G. Trujillo, D. Perez-Staples, 'Male and female condition influence mating performance and sexual receptivity in two tropical fruit flies (Diptera: Tephritidae) with contrasting life histories', *Journal of Insect Physiology*, 2009, 55: 1091–8.

Lisa Locatello, Federica Poli, Maria B. Rasotto, 'Tactic-specific differences in seminal fluid influence sperm performance', *Proceedings of the Royal Society B*, 2013 280: 20122891.

Jin-Feng Yu, Cong Li, Jin Xu, Jian-Hong Liu, Hui Ye, 'Male accessory gland secretions modulate female post-mating behavior in the moth *Spodoptera litura*', *Journal of Insect Behaviour*, 2014, 27: 105–16.

May the best sperm win

Morgan Pearcy, Noémie Delescaille, Pascale Lybaert, Serge Aron, 'Team swimming in ant spermatozoa', *Biology Letters*, 2014, 10: 20140308.

Z. Valentina Zizzari, Nico M. van Straalen, Jacintha Ellers, 'Male–male competition leads to less abundant but more attractive sperm', *Biology Letters*, 2013, 9: 20130762.

Stefan Lüpold, Tim R. Birkhead, David F. Westneat, 'Seasonal variation in ejaculate traits of male red-winged blackbirds (*Agelaius phoeniceus*)', *Behavioral Ecology and Sociobiology*, 2012, 66: 1607–17.

Peter Michalik, Martin J. Ramirez, Christian S. Wirkner, Elisabeth Lipke, 'Morphological evidence for limited sperm production in the enigmatic Tasmanian cave spider *Hickmania troglodytes* (Austrochilidae, Araneae)', *Invertebrate Biology*, 2014, 133 (2): 180–7.

John L. Fitzpatrick, Robert Montgomerie, Julie K. Desjardins, Kelly A. Stiver, Niclas Kolm, Sigal Balshine, 'Female promiscuity promotes the evolution of faster sperm in cichlid fishes', *PNAS*, 27 January 2009, 106 (4): 1128–32.

Martin A. Dziminski, J. Dale Roberts, Maxine Beveridge, Leigh W. Simmons, 'Sperm competitiveness in frogs: slowand steady wins the race', *Proceedings of the Royal Society B*, 2009, 276: 3955–61.

Simone Immler, Sarah R. Pryke, Tim R. Birkhead, Simon C. Griffith, 'Pronounced within-individual plasticity in sperm morphometry across social environments', *Evolution*, 2010, 64 (6): 1634–43.

M. Stürup, B. Baer-Imhoof, D. R. Nash, J.J. Boomsma, B. Baerb, 'When every sperm counts: factors affecting male fertility in the honeybee *Apis mellifera*', *Behavioral Ecology*, 2013, 24 (5): 1192–8.

Sara Calhim, Michael C. Double, Nicolas Margraf, Tim R. Birkhead, Andrew Cockburn, 'Maintenance of sperm variation in a highly promiscuous wild bird', *PLOS ONE*, December 2011, 6 (12): e28809.

Jonathan P. Evans, Patrice Rosengrave, Clelia Gasparini, Neil J. Gemmell, 'Delineating the roles of males and females in sperm competition', *Proceedings of the Royal Society B*, 2013, 280: 20132047.

Montserrat Gomendio, Eduardo R. S. Roldan, 'Implications of diversity in sperm size and function for sperm competition and fertility', *International Journal of Developmental Biology*, 2008, 52: 439–47.

Giulia Cutuli, Stefano Cannicci, Marco Vannini, Sara Fratini, 'Influence of mating order on courtship displays and stored sperm utilization in Hermann's tortoises (*Testudo hermanni hermanni*)', *Behavioral Ecology and Sociobiology*, 2013, 67: 273–81.

Teri J. Orr, Marlene Zuk, 'Does delayed fertilization facilitate sperm competition in bats?', *Behavioral Ecology and Sociobiology*, 2013, 67: 1903–13.

The last laugh: cryptic female choice

Sonja H. Sbilordo, Oliver Y. Martin, 'Pre- and postcopulatory sexual selection act in concert to determine male reproductive success in *Tribolium castaneum*', *Biological Journal of the Linnean Society*, 2014, 112: 67–75.

John R. LaBrecque, Yvette R. Alva-Campbell, Sophie Archambeault, Karen D. Crow, 'Multiple paternity is a shared reproductive strategy in the live-bearing surfperches (Embiotocidae) that may be associated with female fitness', *Ecology and Evolution*, 2014, 4 (12): 2316–29.

Sarah E. Yeates, Sian E. Diamond, Sigurd Einum, Brent C. Emerson, William V. Holt, Matthew J. G. Gage, 'Cryptic choice of conspecific sperm controlled by the impact of ovarian fluid on sperm swimming behavior', *Evolution*, 2013, 67 (12): 3523–56.

Hanne Løvlie, Mark A. F. Gillingham, Kirsty Worley, Tommaso Pizzari, David S. Richardson, 'Cryptic female choice favours sperm from major histocompatibility complex-dissimilar males', *Proceedings of the Royal Society B*, 2013, 280: 20131296.

L. M. King, J. P. Brillard, W. M. Garrett, M. R. Bakst, A. M. Donoghue, 'Segregation of spermatozoa within sperm storage tubules of fowl and turkey hens', *Reproduction*, 2002, 123: 79–86.

Mathew Oliver, Jonathan P. Evans, 'Chemically moderated gamete preferences predict offspring fitness in a broadcast spawning invertebrate', *Proceedings of the Royal Society B*, 2014, 281: 20140148.

Patricia L. R. Brennan, Richard O. Prum, 'The limits of sexual conflict in the narrow sense: new insights from waterfowl biology', *Philosphical Transactions of the Royal Society B*, 2012, 367: 2324–38.

Sarah M. Stai, William A. Searcy, 'Passive Sperm Loss and Patterns of Sperm Precedence in Muscovy Ducks (*Cairina moschata*)', *The Auk*, 2010, 127 (3) :495–502.

Yoshitaka Kamimura, 'Pre- and postcopulatory sexual selection and the evolution of sexually dimorphic traits in earwigs (Dermaptera)', *Entomological Science*, 2014, 17: 139–66.

Christopher R. Friesen, Robert T. Mason, Stevan J. Arnold, Suzanne Estes, 'Patterns of sperm use in two populations of Red-sided Garter Snake (Thamnophis sirtalis parietalis) with long-term female sperm storage', *Canadian Journal of Zoology*, 2014, 92: 33–40.

Claudia C. Buser, Paul I. Ward, Luc F. Bussiere, 'Adaptive maternal plasticity in response to perceptions of larval competition', *Functional Ecology*, 2014, 28: 669–81.

Renée C. Firman, Leigh W. Simmons, 'Sperm competition risk generates phenotypic plasticity in ovum fertilizability', *Proceedings of the Royal Society B*, 2013, 280: 20132097.

Noriyosi Sato, Takashi Kasugai, Hiroyuki Munehara, 'Female pygmy squid cryptically favour small males and fast copulation as observed by removal of spermatangia', *Evolutionary Biology*, 2014, 41: 221–8.

Rulon W. Clark, Gordon W. Schuett, Roger A. Repp, Melissa Amarello, Charles F. Smith, Hans-Werner Herrmann, 'Mating systems, reproductive success, and sexual selection in secretive species: a case study of the western diamond-backed rattlesnake, *Crotalus atrox*', *PLOS ONE*, March 2014, 9 (3): e90616.

Jonathan P. Evans, Patrice Rosengrave, Clelia Gasparini and Neil J. Gemmell, 'Delineating the roles of males and females in sperm competition', *Proceedings of the Royal Society B*, 2013, 280: 20132047.

José Beirão, Craig F. Purchase, Brendan F. Wringe, Ian A. Fleming, 'Wild Atlantic cod sperm motility is negatively affected by ovarian fluid of farmed females', *Aquacult Environ Interact*, 2014, 5: 61–70.

When a boy is a girl too

J. Matthew Hoch, 'Adaptive plasticity of the penis in a simultaneous hermaphrodite', *Evolution*, 2009, 63 (8): 1946–53.

Janet L. Leonard, Jane A. Westfall, John S. Pearse, 'Phally polymorphism and reproductive biology in *Ariolimax* (Ariolimax) *buttoni* (Stylommatophora: Arionidae)', *American Malacological Bulletin*, 2007, 23 (1):121–35.

Yoichi Yusa, Mayuko Takemura, Kota Sawada, Sachi Yamaguchi, 'Diverse, Continuous, and Plastic Sexual Systems in Barnacles', *Integrative and Comparative Biology*, 53 (4): 701–12.

Joris M Koene, Hinrich Schulenburg, 'Shooting darts: co-evolution and counter-adaptation in hermaphroditic snails', *BMC Evolutionary Biology*, 2005, 5: 25.

Ayami Sekizawa, Satoko Seki, Masakazu Tokuzato, Sakiko Shiga, Yasuhiro Nakashima, 'Disposable penis and its replenishment in a simultaneous hermaphrodite', *Biology Letters*, 2013, 9: 20121150.

Joris M. Koene, Tina Pfortner, Nico K. Michiels, 'Piercing the partner's skin influences sperm uptake in the earthworm *Lumbricus terrestris*', *Behavioral Ecology and Sociobiology*, 2005, 59: 243–9.

Lukas Schärer, 'Evolution: Don't be so butch, dear!', *Current Biology*, 2014, 24 (8): R311–13.

Lucas Marie-Orleacha, Tim Janicke, Lukas Schärer, 'Effects of mating status on copulatory and postcopulatory behaviour in a simultaneous hermaphrodite', *Animal Behaviour*, 2013, 85: 453–61.

Tim Janicke, Lucas Marie-Orleach, Katrien De Mulder, Eugene Berezikov, Peter Ladurner, Dita B. Vizoso, Lukas Schärer, 'Sex allocation adjustment to mating group size in a simultaneous hermaphrodite', *Evolution*, 2013, 67 (11): 3233–42.

M. K. Hart, A. Svoboda, D. Mancilla Cortez, 'Phenotypic plasticity in sex allocation for a simultaneously hermaphroditic coral reef fish', *Coral Reefs*, 2011, 30: 543–8.

Yumi Nakadera, Joris M. Koene, 'Reproductive strategies in hermaphroditic gastropods: conceptual and empirical approaches', *Canadian Journal of Zoology*, 91: 367–81.

D. Schleicherová, G. Sella, S. Meconcelli, R. Simonini, M. P. Martino, P. Cervella, M.C. Lorenzi, 'Does the cost of a function affect its degree of plasticity? A test on plastic sex allocation in three closely related species of hermaphrodites', *Journal of Experimental Marine Biology and Ecology*, 2014, 453: 148–53.

Nils Anthes, Nico K. Michiels, 'Precopulatory stabbing, hypodermic injections and unilateral copulations in a hermaphroditic sea slug', *Biology Letters*, 2007, 3: 121–4.

Joris M. Koene, Thor-Seng Liew, Kora Montagne-Wajer, Menno Schilthuizen, 'A syringe-like love dart injects male accessory gland products in a tropical hermaphrodite', *PLOS ONE*, July 2013, 8 (7): e69968.

Kazuki Kimura, Kaito Shibuya, Satoshi Chiba, 'The mucus of a land snail love-dart suppresses subsequent matings in darted individuals', *Animal Behaviour*, 2013, 85: 631–5.

Lukas Schärer, Ido Pen, 'Sex allocation and investment into pre- and post-copulatory traits in simultaneous hermaphrodites: the role of polyandry and local sperm competition', *Philosphical Transactions of the Royal Society B*, 2013, 368: 20120052.

Rolanda Lange, Tobias Gerlach, Joscha Beninde, Johanna Werminghausen, Verena Reichel, Nils Anthes, 'Female fitness optimum at intermediate mating rates under traumatic mating', *PLOS ONE*, August 2012, 7 (8): e43234.

Yumi Nakadera, Elferra M. Swart, Jeroen N.A. Hoffer, Onno den Boon, Jacintha Ellers, Joris M. Koene, 'Receipt of seminal fluic proteins causes reduction of male investment in a simultaneous hermaphrodite', *Current Biology*, 2014, 24: 859–62.

N. K. Michiels, L. J. Newman, 'Sex and violence in hermaphrodites', *Nature*, 12 February 1998, 391: 647.

Janet L. Leonard, 'Sexual selection and hermaphroditic organisms: Testing theory', *Current Zoology*, 2013, 59 (4): 579–88.

Lukas Schärer, D. Timothy J. Littlewood, Andrea Waeschenbach, Wataru Yoshida, Dita B. Vizoso, 'Mating behavior and the evolution of sperm design', *PNAS*, 25 January 2011, 108 (4): 1490–95.

Tim Janicke, Halil Kesselring, Lukas Schärer, 'Strategic mating effort in a simultaneous hermaphrodite: The role of the partner's feeding status', *Behavioral Ecology and Sociobiology*, 2012, 66: 593–601.

Sex thy self

Bruce Bagemihl, *Biological Exuberance: Animal Homosexuality and Natural Diversity*, Stonewall Inn Editions, 2000.

Martin Wikelski, Silke Baurle, 'Pre-copulatory ejaculation solves time constraints during copulations in marine iguanas', *Proceedings of the Royal Society B*, 1996, 263(1369): 439–444.

Jane M. Waterman, 'The adaptive function of masturbation in a promiscuous African ground squirrel', *PLOS ONE*, September 2010, 5 (9): e13060.

S. M. McDonnell, M. Henry, F. Bristoif, 'Spontaneous erection and masturbation in equids', *Journals of Reproduction and Fertility*, 1991, Suppl. 44: 664–5.

Constance Dubuc, Sean P. Coyne, Dario Maestripieri, 'Effect of mating activity and dominance rank on male masturbation among free-ranging male rhesus macaques', *Ethology*, 2013, 119: 1006–1013.

John L. Hoogland, 'Estrus and copulation of Gunnison's prairie dogs', *Journal of Mammalogy*, 1998, 79 (3): 887–97.

Benjamin L. Hart, Elizabeth Korinek, Patricia Brennan, 'Postcopulatory genital grooming in male rats: prevention of sexually transmitted infections', *Physiology & Behavior*, 1987, 41: 321–5.

Sue M. McDonnell, Amy L. Hinze, 'Aversive conditioning of periodic spontaneous erection adversely affects sexual behaviour and semen in stallions', *Animal Reproduction Science*, 2005, 89: 77–92.

Klaus Reinhardt, Michael T. Siva-Jothy, 'An advantage for young sperm in the house cricket *Acheta domesticus*', *The American Naturalist*, June 2005, 165 (6): 718–23.

The big O

Brendan P. Zietsch, Pekka Santtila, 'Genetic analysis of orgasmic function in twins and siblings does not support the by-product theory of female orgasm', *Animal Behaviour*, 2011, 82: 1097–1101.

Emile van Lieshout, 'Male genital length and mating status differentially affect mating behaviour in an earwig', *Behavioral Ecology and Sociobiology*, 2011, 65: 149–56.

David A. Puts, Khytam Dawood, Lisa L. M. Welling, 'Why women have orgasms: an evolutionary analysis', *Archives of Sexual Behavior*, 2012, 41: 1127–43.

Robert King, Jay Belsky, 'A typological approach to testing the evolutionary functions of human female orgasm', *Archives of Sexual Behavior*, 2012, 41: 1145–60.

Roy J. Levin, 'The pharmacology of the human female orgasm – its biological and physiological backgrounds', *Pharmacology, Biochemistry and Behavior*, 2014, 121: 62–70.

Alfonso Troisi, Monica Carosi, 'Female orgasm rate increases with male dominance in Japanese macaques', *Animal Behaviour*, 1998, 56: 1261–6.

M. Winterbottom, T. Burke, T. R. Birkhead, 'A stimulatory phalloid organ in a weaver bird', *Nature*, 6 May 1999, 399: 28–9.

M. Winterbottom, T. Burke, T. R. Birkhead, 'The phalloid organ, orgasm and sperm competition in a polygynandrous bird: the red-billed buffalo weaver (*Bubalornis niger*)', *Behavioral Ecology and Sociobiology*, 2001, 50: 474–82.

Bestiality

Daisuke Kyogoku, Takayoshi Nishida, 'The mechanism of the fecundity reduction in *Callosobruchus maculatus* caused by *Callosobruchus chinensis* males', *Population Ecology*, 2013, 55: 87–93.

P. J. Nico de Bruyn, Cheryl A. Tosh, Marthan N. Bester, 'Sexual harassment of a king penguin by an Antarctic fur seal', *Journal of Ethology*, 2008, 26: 295–7.

John F. Benson, Brent R. Patterson, Peter J. Mahoney, 'A protected area influences genotype-specific survival and the structure of a Canis hybrid zone', *Ecology*, 2014, 95(2): 254–64.

Grava, T. Grava, R. Didier, L. A. Lait, J. Dosso, E. Koran, T.M. Burg, K. A. Otter, 'Interspecific dominance relationships and hybridization between black-capped and mountain chickadees', *Behavioral Ecology*, 2012, 23: 566–72.

Yoshitaka Kamimura, 'Correlated evolutionary changes in *Drosophila* female genitalia reduce the possible infection risk caused by male copulatory wounding', *Behavioral Ecology and Sociobiology*, 2012, 66: 1107–14.

Henrique Caldeira Costa, Emanuel Teixeira da Silva, Pollyanna Silva Campos, Marina Paula da Cunha Oliveira, André Valle Nunes, Patrícia da Silva Santos, 'The Corpse Bride: a case of Davian Behaviour in the Green Ameiva (Ameiva *ameiva*) in southeastern Brazil', *Herpetology Notes*, 2010, 3: 79–83.

T. J. Izzo, D. J. Rodrigues, M. Menin, A. P. Lima, W. E. Magnusson, 'Functional necrophilia: a profitable anuran reproductive strategy?', *Journal of Natural History*, 2012, 46: 47–48.

S. Piraino, D. De Vito, J. Bouillon, F. Boero, 'Larval necrophilia: the odd life cycle of a pandeid hydrozoan in the Weddell Sea shelf', *Polar Biology*, 2003, 26: 178–85.

Douglas G. D. Russell, William J. L. Sladen, David G. Ainley, 'Dr George Murray Levick (1876–1956): unpublished notes on the sexual habits of the Adélie penguin', *Polar Record*, 2012, 48 (247): 387–93.

Lucas Bezerra de Mattos Brito, Igor Roberto Joventino, Samuel Cardozo Ribeiro, Paulo Cascon, 'Necrophiliac behavior in the "cururu" toad, *Rhinella jimi* Steuvax, 2002, (Anura, Bufonidae) from Northeastern Brazil', *North-Western Journal of Zoology*, 2012, 8 (2): 365–6.

R. H. Andrews, T. N. Petney, C. M. Bull, 'Reproductive interference between three parapatric species of reptile tick', *Oecologia*, 1982, 52 (2): 281–6.

Kazunori Yoshizawa, Rodrigo L. Ferreira, Yoshitaka Kamimura, Charles Lienhard, 'Female penis, male vagina, and their correlated evolution in a cave insect', *Current Biology*, 2014, 24: 1006–10.

Teiji Sota, Kohei Kubota, 'Genital lock-and-key as a selective agent against hybridization', *Evolution*, 1998, 52 (5): 1507–13.

Alejandra Valero, Constantino Macias Garcia, Anne E. Magurran, 'Heterospecific harassment of native endangered fishes by invasive guppies in Mexico', *Biology Letters*, 2008, 4: 149–52.

Erica L. Larson, Thomas A. White, Charles L. Ross, Richard G. Harrison, 'Gene flow and the maintenance of species boundaries', *Molecular Ecology*, 2014, 23: 1668–78.

Pamela M. Willis, Michael J. Ryan, Gil G. Rosenthal, 'Encounter rates with conspecific males influence female mate choice in a naturally hybridizing fish', *Behavioral Ecology*, 2011, 22: 1234–40.

Denson Kelly McLain, Donald J. Shure, 'Pseudocompetition: interspecific displacement of insect species through misdirected courtship', *Oikos*, July 1987, 49 (3): 291–6.

Nikolai J. Tatarnic, Gerasimos Cassis, 'Surviving in sympatry: paragenital divergence and sexual mimicry between a pair of traumatically inseminating plant bugs', *The American Naturalist*, October 2013, 182 (4): 542–51.

William A. Haddad, Ryan R. Reisinger, Tristan Scott, Martha N. Bester, P. J. Nico de Bruyn, 'Multiple occurrences of king penguin (*Aptenodytes patagonicus*) sexual harassment by Antarctic fur seals (*Arctocephalus gazella*), *Polar Biology*, November 2014.

Axel Hochkirch, Julia Gröning, Amelie Bücker, 'Sympatry with the devil: reproductive interference could hamper species coexistence', *Journal of Animal Ecology*, July 2007, 76 (4): 633–42.

Emily R. Burdfield-Steel, David M. Shuker, 'Reproductive interference', *Current Biology*, 21 (12) R450–1.

Attila Hettyey, Peter B. Pearman, 'Social environment and reproductive interference affect reproductive success in the frog *Rana latastei*', *Behavioral Ecology*, 2003, 14: 294–300.

Julia Gröning, Axel Hochkirch, 'Reproductive interference between animal species', *The Quarterly Review of Biology*, September 2008, 83 (3): 257–82.

Mara Casalini, Martin Reichard, André Phillips, Carl Smith, 'Male choice of mates and mating resources in the rose bitterling (*Rhodeus ocellatus*)', *Behavioral Ecology*, 2013, 24(5): 1199–1204.

Jean Secondi, Marc Théry, 'An ultraviolet signal generates a conflict between sexual selection and species recognition in a newt', *Behavioral Ecology and Sociobiology*, 2014, 68: 1049–58.

Queer as fauna

Nathan W. Bailey, Marlene Zuk, 'Same-sex sexual behaviour and evolution', *Trends in Ecology and Evolution*, 2009, 24 (8): 439–46.

Nathan W. Bailey, Nicholas French, 'Same-sex sexual behaviour and mistaken identity in male field crickets, *Teleogryllus oceanicus*', *Animal Behaviour*, 2012, 84: 1031–8.

K. E. Levan, T. Y. Fedina, S. M. Lewis, 'Testing multiple hypotheses for the maintenance of male homosexual copulatory behaviour in flour beetles', *Journal of Evolutionary Biology*, 2009, 22: 60–70.

Lindsay C. Young, Eric A. VanderWerf, 'Adaptive value of same-sex pairing in Laysan albatross', *Proceedings of the Royal Society B*, 2014, 281: 20132473.

Giovanni Benelli, Gabriele Gennari, Alessandra Francini, Angelo Canale, 'Longevity costs of same-sex interactions: first evidence from a parasitic wasp', *Invertebrate Biology*, 2013, 132(2): 156–62.

Nathan W. Bailey, Jessica L. Hoskins, Jade Green, Michael G. Ritchie, 'Measuring same-sex sexual behaviour: the influence of the male social environment', *Animal Behaviour*, 2013, 86: 91–100.

Xueyan Yang, Isabelle Attane, Shuzhuo Li, Qunlin Zhang, 'On same-sex behaviors among male bachelors in rural China: evidence from a female shortage context', *American Journal of Men's Health*, 2012, 6: 108.

Puya Abbassi, Nancy Tyler Burley, 'Nice guys finish last: same-sex sexual behaviour and pairing success in male budgerigars', *Behavioral Ecology*, July/August 2012, 23 (4): 775.

Inon Scharf, Oliver Y. Martin, 'Same-sex sexual behavior in insects and arachnids: prevalence, causes, and consequences', *Behavioral Ecology and Sociobiology*, 2013, 67: 1719–30.

Biljana Stojkovic, Darka Seslija Jovanovic, Branka Tucic, Nikola Tucic, 'Homosexual behaviour and its longevity cost in females and males of the seed beetle *Acanthoscelides obtectus*', *Physiological Entomology*, 2010, 35: 308–16.

Sexual coercion

Martin N. Muller, Melissa Emery Thompson, Sonya M. Kahlenberg, Richard W. Wrangham, 'Sexual coercion by male chimpanzees shows that female choice may be more apparent than real', *Behavioral Ecology and Sociobiology*, 2011, 65: 921–33.

Pablo Polo, Victoria Hernandez-Lloreda, Fernando Colmenares, 'Male takeovers are reproductively costly to females in Hamadryas baboons: a test of the sexual coercion hypothesis', *PLOS ONE*, March 2014, 9(3): e90996.

P. M. R. Clarke, J. E. B. Halliday, L. Barrett, S. P. Henzi, 'Chacma baboon mating markets: competitor suppression mediates the potential for intersexual exchange', *Behavioral Ecology*, 2010, 21: 1211–20.

Martin N. Muller, Sonya M. Kahlenberg, Melissa Emery Thompson, Richard W. Wrangham, 'Male coercion and the costs of promiscuous mating for female chimpanzees', *Proceedings of the Royal Society B*, 2007, 274: 1009–14.

Martin Surbeck, Gottfried Hohmann, 'Intersexual dominance relationships and the influence leverage on the outcome of conflicts in wild bonobos (*Pan paniscus*)', *Behavioral Ecology and Sociobiology*, 2013, 67: 1767–80.

Maria A. van Noordwijk, Carel P. van Schaik, 'Intersexual food transfer among orang-utans: do females test males for coercive tendency?', *Behavioral Ecology and Sociobiology*, 2009, 63: 883–90.

Larissa Swedell, Liane Leedom, Julian Saunders, Mathew Pines, 'Sexual conflict in a polygynous primate: costs and benefits of a male-imposed mating system', *Behavioral Ecology and Sociobiology*, 2014, 68: 263–73.

Holly A. MacCormick, Daniel R. MacNulty, Anna L. Bosacker, Clarence Lehman, Andrea Bailey, D. Anthony Collins, Craig Packer, 'Male and female aggression: lessons from sex, rank, age, and injury in olive baboons', *Behavioral Ecology*, 2012, 23: 684–91.

Cheryl Denise Knott, Melissa Emery Thompson, Rebecca M. Stumpf, Matthew H. McIntyre, 'Female reproductive strategies in orangutans, evidence for female choice and counterstrategies to infanticide in a species with frequent sexual coercion', *Proceedings of the Royal Society B*, 2010, 277: 105–113.

F. Galimberti, L. Boitani, I. Marzetti, 'Female strategies of harassment reduction in southern elephant seals', *Ethology, Ecology & Evolution*, 2000, 12 (4): 367–88.

Elise Huchard, Cindy I. Canale, Chloé Le Gros, Martine Perret, Pierre-Yves Henry, Peter M. Kappeler, 'Convenience polyandry or convenience polygyny? Costly sex under female control in a promiscuous primate', *Proceedings of the Royal Society B*, 2012, 279: 1371–79.

P. J. N. de Bruyn, C. A. Tosh, M. N. Bester, E. Z. Cameron, T. McIntyre, I. S. Wilkinson, 'Sex at sea: alternative mating system in an extremely polygynous mammal', *Animal Behaviour*, 2011, 82: 445–51.

Cannibalistic females and the males that love them

Maydianne C. B. Andrade, Lei Gu, Jeffrey A. Stoltz, 'Novel male trait prolongs survival in suicidal mating', *Biology Letters*, 2005, 1: 276–9.

Simona Kralj-Fiser, Jutta M. Schneider, Matjaz Kuntner, 'Challenging the aggressive spillover hypothesis: is pre-copulatory sexual cannibalism a part of a behavioural syndrome?', *Ethology*, 2013, 119: 615–23.

Jonathan N. Pruitt, Carl N. Keiser, 'The Aggressive Spillover Hypothesis: Existing Ailments and Putative Remedies', *Ethology*, 2013, 119: 807–10.

Simona Kralj-Fišer, Graciela A. Sanguino Mostajo, Onno Preik, Stano Pekár, Jutta M. Schneider, 'Assortative mating by aggressiveness type in orb weaving spiders', *Behavioral Ecology*, 2013, 24 (4), 824–31.

Trine Bilde, Cristina Tuni, Rehab Elsayed, Stano Pekár, Søren Toft, 'Death feigning in the face of sexual cannibalism', *Biology Letters*, 2006, 2: 23–5.

Murray P. Fea, Margaret C. Stanley, Gregory I. Holwell, 'Fatal attraction: sexually cannibalistic invaders attract naive native mantids', *Biology Letters*, 2013, 9: 20130746.

Lutz Fromhage, Jutta M. Schneider, 'Safer sex with feeding females: sexual conflict in a cannibalistic spider', *Behavioral Ecology*, 2005, 16: 377–82.

Stefan H. Nessler, Gabriele Uhl, Jutta M. Schneider, 'Sexual cannibalism facilitates genital damage in *Argiope lobata* (Araneae: Araneidae)', *Behavioral Ecology and Sociobiology*, September 2009, 63: 355–62.

Ruben Rabaneda-Bueno, Sara Aguado, Carmen Fernandez-Montraveta, Jordi Moya-Laraño, 'Does Female Personality Determine Mate Choice Through Sexual Cannibalism?', *Ethology*, 2014, 120: 238–48.

Jonathan N. Pruitt, Aric W. Berning, Brian Cusack, Taylor A. Shearer, Mathew McGuirk, Anna Coleman, Robin Y. Y. Eng, Fawn Armagost, Kayla Sweeney, Nishant Singh, 'Precopulatory sexual cannibalism causes increase egg case production, hatching success, and female attractiveness to males', *Ethology*, 2014, 120: 453–62.

Daiqin Li, Joelyn Oh, Simona Kralj-Fiser, Matjaz Kuntner, 'Remote copulation: male adaptation to female cannibalism', *Biology Letters*, 2012, 8: 512–15.

Anita Aisenberg, Fernando G. Costa, Macarena González, 'Male sexual cannibalism in a sand-dwelling wolf spider with sex role reversal', *Biological Journal of the Linnean Society*, 2011, 103: 68–75.

Lenka Sentenska, Stano Pekar, 'Eat or not to eat: reversed sexual cannibalism as a male foraging strategy in the spider *Micaria sociabilis* (Araneae: Gnaphosidae)', *Ethology*, 2014, 120: 511–18.

William D. Brown, Gregory A. Muntz, Alexander J. Ladowski, 'Low mate encounter rate increases male risk taking in a sexually cannibalistic praying mantis', *PLOS ONE*, April 2012, 7(4): e35377.

Katherine L. Barry, Gregory I. Holwell, Marie E. Herberstein, 'Male mating behaviour reduces the risk of sexual cannibalism in an Australian praying mantid', *Journal of Ethology*, 2009, 27: 377–83.

Stefan H. Nessler, Gabriele Uhl, Jutta M. Schneider, 'Scent of a woman – the effect of female presence on sexual cannibalism in an orb-weaving spider (Araneae: Araneidae)', *Ethology*, 2009, 115: 633–40.

Hiroshi Watanabe, Eizi Yano, 'Behavioral response of male mantid *Tenodera aridifolia* (Mantodea: Mantidae) to windy conditions as a female approach strategy', *Entomological Science*, 2012, 15: 384–91.

Klaas W. Welke, Jutta M. Schneider, 'Males of the orb-web spider *Argiope bruennichi* sacrifice themselves to unrelated females', *Biology Letters*, 2010, 6: 585–88.

Line Spinner Hansen, Sofia Fernandez Gonzalez, Søren Toft, Trine Bilde, 'Thanatosis as an adaptive male mating strategy in the nuptial gift-giving spider *Pisaura mirabilis*', *Behavioral Ecology*, 2008, 19: 546–51.

Elina Immonen, Anneli Hoikkala, Anahita J. N. Kazem, Michael G. Ritchie, 'When are vomiting males attractive? Sexual selection on condition-dependent nuptial feeding in Drosophila *subobscura*', *Behavioral Ecology*, 2009, 20: 289–95.

J. Chadwick Johnson, Patricia Trubl, Valerie Blackmore, Lindsay Miles, 'Male black widows court well-fed females more than starved females: silken cues indicate sexual cannibalism risk', *Animal Behaviour*, 2011, 82: 383–90.

Chastity belts

Qi Qi Lee, Joelyn Oh, Simona Kralj-Fiser, Matjaz Kuntner, Daiqin Li, 'Emasculation: gloves-off strategy enhances eunuch spider endurance', *Biology Letters*, 2012, 8: 733–5.

Simona Kralj-Fiser, Matjaz Gregoric, Shichang Zhang, Daiqin Li, Matjaz Kuntner, 'Eunuchs are better fighters', *Animal Behaviour*, 2011, 81: 933–9.

Anita Aisenberg, William G. Eberhard, 'Female cooperation in plug formation in a spider: effects of male copulatory courtship', *Behavioral Ecology*, 2009, 20: 1236–41.

Christopher R. Friesen, Richard Shine, Randolph W. Krohmer, Robert T. Mason, 'Not just a chastity belt: the functional significance of mating plugs in garter snakes, revisited', *Biological Journal of the Linnean Society*, 2013, 109: 893–907.

Christopher R. Friesen, Emily J. Uhrig, Mattie K. Squire, Robert T. Mason, Patricia L. R. Brennan, 'Sexual conflict over mating in red-sided garter snakes (*Thamnophis sirtalis*) as indicated by experimental manipulation of genitalia', *Proceedings of the Royal Society B*, 2014, 281: 20132694.

M. E. Herberstein, A. E. Wignall, S. H. Nessler, A. M. T. Harmer, J. M. Schneider, 'How effective and persistent are fragments of male genitalia as mating plugs?', *Behavioral Ecology*, 2012, 23: 1140–5.

Matjaz Kuntner, Matjaz Gregoric, Shichang Zhang, Simona Kralj-Fiser, Daiqin Li, 'Mating plugs in polyandrous giants: which sex produces them, when, how and why?', *PLOS ONE*, July 2012, 7(7): e40939.

Joyce A. Parga, 'Male Post-Ejaculatory Mounting in the Ring-Tailed Lemur (*Lemur catta*): A Behavior Solicited by Females?', *Ethology*, 2010, 116: 832–42.

Renée C. Firman, 'Female fitness, sperm traits and patterns of paternity in an Australian polyandrous mouse', *Behavioral Ecology and Sociobiology*, 2014, 68: 283–90.

Katrin Kunz, Melanie Witthuhn, Gabriele Uhl, 'Do the size and age of mating plugs alter their efficacy in protecting paternity?', *Behavioral Ecology and Sociobiology*, 2014, 68: 1321–8.

Lutz Fromhage, 'Mating unplugged: a model for the evolution of mating plug (dis)placement', *Evolution*, 2011, 66 (1): 31–9.

Steven K. Schwartz, William E. Wagner, Jr, Eileen A. Hebets, 'Obligate male death and sexual cannibalism in dark fishing spiders', *Animal Behaviour*, 2014, 93: 151–6.

José L. Tlachi-López, Jose R. Eguibar, Alonso Fernández-Guasti, Rosa Angélica Lucio, 'Copulation and ejaculation in male rats under sexual satiety and the Coolidge effect', *Physiology & Behavior*, 2012, 106: 626–30.

Sarah Althaus, Alain Jacob, Werner Graber, Deborah Hofer, Wolfgang Nentwig, Christian Kropf, 'A double role of sperm in scorpions: the mating plug of *Euscorpius italicus* (Scorpiones: Euscorpiidae) consists of sperm', *Journal of Morphology*, 2010, 271: 383–93.

A. Lucio, Rodriguez-Piedracruz, L. Tlachi-Lopez, Garcia-Lorenzana Fernandez-Guasti, 'Copulation without seminal expulsion: the consequence of sexual satiation and the Coolidge effect', *Andrology*, 2014, 2: 450–7.

Karen E. Munroe, John L. Koprowski, 'Copulatory plugs of round-tailed ground squirrels (*Xerospermophilus tereticaudus*)', *Southwestern Naturalist*, 2012, 57(2): 208–10.

Anita Aisenberg, Gilbert Barrantes, 'Sexual behavior, cannibalism, and mating plugs as sticky traps in the orb weaver spider *Leucauge argyra* (Tetragnathidae)', *Naturwissenschaften*, 2011, 98: 605–13.

Menno Schilthuizen, *Nature's Nether Regions*, Viking 2014.

Fifty shades of BDSM in the animal kingdom

Klaus Reinhardt, Ewan Harney, Richard Naylor, Stanislav Gorb, Michael T. Siva-Jothy, 'Female-limited polymorphism in the copulatory organ of a traumatically inseminating insect', *The American Naturalist*, December 2007, 170 (6): 931–5.

Rolanda Lange, Johanna Werminghausen, Nils Anthes, 'Cephalo-traumatic secretion transfer in a hermaphrodite sea slug', *Proceedings of the Royal Society B*, 2014, 281: 20132424.

Joris M. Koene, Tina Pförtner, Nico K. Michiels, 'Piercing the partner's skin influences sperm uptake', *Behavioral Ecology and Sociobiology*, 2005, 59: 243–9.

Casie Horgan, Gretel Terrero, Gary Wessel, 'In the extremes – traumatic insemination', *Molecular Reproduction and Development*, 2011, 78: 5.

Weihao Zhong, Colin D. McClure, Cara R. Evans, David T. Mlynski, Elina Immonen, Michael G. Ritchie, Nicholas K. Priest, 'Immune anticipation of mating in *Drosophila: Turandot M* promotes immunity against sexually transmitted fungal infections', *Proceedings of the Royal Society B*, 2013, 280: 20132018.

Christopher R. Friesen, Emily J. Uhrig, Mattie K. Squire, Robert T. Mason and Patricia L. R. Brennan, 'Sexual conflict over mating in red-sided garter snakes (*Thamnophis sirtalis*) as indicated by experimental manipulation of genitalia', *Proceedings of the Royal Society B*, 2014, 281: 20132694.

Milan Rezac, 'The spider Harpactea sadistica: co-evolution of traumatic insemination and complex female genital morphology in spiders', *Proceedings of the Royal Society B*, 2009, 276: 2697–2701.

Camilla Ryne, 'Homosexual interactions in bed bugs: alarm pheromones as male recognition signals', *Animal Behaviour*, 2009, 78: 1471–5.

Yutaka Okuzaki, Yasuoki Takami, Yuzo Tsuchiya, Teiji Sota, 'Mating Behavior and the Function of the Male Genital Spine in the Ground Beetle *Carabus clathratus*', *Zoological Science*, 2012, 29(7): 428–32.

Amy Backhouse, Steven M. Sait, Tom C. Cameron, 'Multiple mating in the traumatically inseminating Warehouse pirate bug, *Xylocoris flavipes:* effects on fecundity and longevity', *Biology Letters*, 2012, 8: 706–9.

Martin Haasel, Anna Karlsson, 'Mating and the inferred function of the genital system of the nudibranch, *Aeolidiella glauca* (Gastropoda: Opisthobranchia: Aeolidioidea)', *Invertebrate Biology*, 2000, 119(3): 287–98.

Stephanie M. Rollmann, Lynne D. Houck, Richard C. Feldhoff, 'Proteinaceous pheromone affecting female receptivity in a terrestrial salamander', *Science*, 1999, 285: 1907.

Bora Inceoglu, Jozsef Lango, Jie Jing, Lili Chen, Fuat Doymaz, Isaac N. Pessah, Bruce D. Hammock, 'One scorpion, two venoms: Prevenom of *Parabuthus transvaalicus* acts as an alternative type of venom with distinct mechanism of action', *PNAS*, February 2003, 100(3): 922–7.

Julianna L. Johns, J. Andrew Roberts, David L. Clark, George W. Uetz, 'Love bites: male fang use during coercive mating in wolf spiders', *Behavioral Ecology and Sociobiology*, 2009, 64: 13–18.

Yoshitaka Kamimura, Hiroyuki Mitsumoto, Chow-Yang Lee, 'Duplicated Female Receptacle Organs for Traumatic Insemination in the Tropical Bed Bug Cimex hemipterus: Adaptive Variation or Malformation?', *PLOS ONE*, February 2014, 9(2): e89265.

Joshua B. Benoit, Andrew J. Jajack, Jay A. Yoder, 'Multiple traumatic insemination events reduce the ability of bed bug females to maintain water balance', *Journal of Comparative Physiology B*, 2012, 182: 189–98.

Rolanda Lange, Tobias Gerlach, Joscha Beninde, Johanna Werminghausen, Verena Reichel, Nils Anthes, 'Female Fitness Optimum at Intermediate Mating Rates under Traumatic Mating', *PLOS ONE*, August 2012, 7(8): e43234.

Rolanda Lange, Klaus Reinhardt, Nico K. Michiels, Nils Anthes, 'Functions, diversity, and evolution of traumatic mating', *Biological Review*, 2013, 88: 585–601.

Nikolai J. Tatarnic, Gerasimos Cassis, Michael T. Siva-Jothy, 'Traumatic insemination in terrestrial arthropods', *Annual Review of Entomology*, 2014, 59: 245–61.

Rolanda Lange, Johanna Werminghausen, Nils Anthes, 'Does traumatic secretion transfer manipulate mating roles or reproductive output in a hermaphroditic sea slug?', *Behavioral Ecology and Sociobiology*, 2013, 67: 1239–47.

Give the penis a bone

Steven H. Ferguson, Serge Larivière, 'Are long penis bones an adaption to high latitude snowy environments?', *Oikos*, 2004, 105: 2.

Steven A. Ramm, 'Sexual selection and genital evolution in mammals: a phylogenetic analysis of baculum length', *The American Naturalist*, March 2007, 169(3): 360–9.

Jean-François Lemaître, Steven A. Ramm, Nicola Jennings, Paula Stockley, 'Genital morphology linked to social status in the bank vole (*Myodes glareolus*)', *Behavioral Ecology and Sociobiology*, 2012, 66: 97–105.

D. J. Hosken, K. E. Jones, K. Chipperfield, A. Dixson, 'Is the bat os penis sexually selected?', *Behavioral Ecology and Sociobiology*, 2012, 66: 97–105.

Leigh W. Simmons, Renee C. Firman, 'Experimental evidence for the evolution of the mammalian baculum by sexual selection', *Evolution*, 2013, 68(1): 276–83.

Paula Stockley, Steven A. Ramm, Amy L. Sherborne, Michael D. F. Thom, Steve Paterson, Jane L. Hurst, 'Baculum morphology predicts reproductive success of male house mice under sexual selection', *BMC Biology*, 2013, 11: 66.

John L. Fitzpatrick, Maria Almbro, Alejandro Gonzalez-Voyer, Niclas Kolm, Leigh W. Simmons, 'Male contest competition and the coevolution of weaponry and testes in pinnipeds', *Evolution*, 2012, 66(11): 3595–3604.

Agata J. Krawczyk, Anna W. Malecha, Piotr Tryjanowski, 'Is baculum size dependent on the condition of males in the polecat *Mustela putorius?*', *Folia Zoologica*, 2011, 60(3): 247–52.

P. Stockley, 'Sperm competition risk and male genital anatomy: comparative evidence for reduced duration of female sexual receptivity in primates with penile spines', *Evolutionary Ecology*, 2002, 16: 123–37.

David J. Yurkowski, Magaly Chambellant, Steven H. Ferguson, 'Bacular and testicular growth and allometry in the ringed seal (Pusa *hispida*): evidence of polygyny?', Journal of Mammalogy, 2011, 92(4): 803–10.

Paula Stockley, 'The baculum', *Current Biology*, 22(24): R1032.

Girls with boy bits

Gerald R. Cunha, Gail Risbridger, Hong Wang, Ned J. Place, Mel Grumbach, Tristan J. Cunha, Mary Weldele, Al J. Conley, Dale Barcellos, Sanjana Agarwal, Argun Bhargava, Christine Drea, Geoffrey L. Hammond, Penti Siiteri, Elizabeth M. Coscia, Michael J. McPhaul, Laurence S. Baskin, Stephen E. Glickman, 'Development of the external genitalia: Perspectives from the spotted hyena (*Crocuta crocuta*)', *Differentiation*, 2014, 87: 4–22.

Marion L. East, Heribert Hofer, Wolfgang Wickler, 'The erect "penis" is a flag of submission in a female-dominated society: greetings in Serengeti spotted hyenas', *Behavioral Ecology and Sociobiology*, 1993, 33: 355–70.

C. M. Drea, N. J. Place, M. L. Weldele, E. M. Coscia, P. Licht, S. E. Glickman, 'Exposure to naturally circulating androgens during foetal life incurs direct reproductive costs in female spotted hyenas, but is prerequisite for male mating', *Proceedings of the Royal Society London B*, 2002, 269: 1981–7.

Martin N. Muller, Richard Wrangham, 'Sexual mimicry in hyenas', *Quarterly Review of Biology*, March 2002, 77(1): 3–16.

Marion L. East, Herbert Hofer, 'The peniform clitoris of female spotted hyaenas', *TREE*, 10 October 1997, 12: 401–2.

Laurence G. Frank, 'Evolution of genital masculinization: why do female hyaenas have such a large "penis"?', *TREE*, 2 February 1997, 12: 58–62.

Christine M. Drea, Anne Weil, 'External genital morphology of the ring-tailed lemur (*Lemur catta*): females are naturally "masculinized"', *Journal of Morphology*, 2008, 269: 451–63.

Stephen E. Glickmana, Roger V. Short, Marilyn B. Renfree, 'Sexual differentiation in three unconventional mammals: Spotted hyenas, elephants and tammar wallabies', *Hormones and Behavior*, 2005, 48: 403–17.

Transvestites

Lindsey Swierk, Tracy Langkilde, 'Bearded ladies: females suffer fitness consequences when bearing male traits', *Biology Letters*, 2013, 9: 20130644.

Audrey Sternalski, François Mougeot, Vincent Bretagnolle, 'Adaptive significance of permanent female mimicry in a bird of prey', *Biology Letters*, 2012, 8: 167–170.

Culum Brown, Martin P. Garwood, Jane E. Williamson, 'It pays to cheat: tactical deception in a cephalopod social signalling system', *Biology Letters*, 2012, 8: 729–32.

Viviana Cadena, 'Bi-gender cross-dressing fools rival suitors', *Journal of Experimental Biology*, 2013, 216(9): iv.

Y. Takahashi, G. Morimoto, M. Watanabe, 'Ontogenetic colour change in females as a function of antiharassment strategy', *Animal Behaviour*, 2012, 84: 685–92.

Tom D. Schultz, Ola M. Fincke, 'Lost in the crowd or hidden in the grass: signal apparency of female polymorphic damselflies in alternative habitats', *Animal Behaviour*, 2013, 86: 923–31.

Yuma Takahashi, Kotaro Kagawa, Erik I. Svensson, Masakado Kawata, 'Evolution of increased phenotypic diversity enhances population performance by reducing sexual harassment in damselflies', *Nature Communications*, 2014, 5: 4468.

Mingzi Xu, Ariana L. Cerreta, Tom D. Schultz, Ola M. Fincke, 'Selective use of multiple cues by males reflects a decision rule for sex discrimination in a sexually mimetic damselfly', *Animal Behaviour*, 2014, 92: 9–18.

Michael J. Ryan, Craig M. Pease, Molly R. Morris, 'A genetic polymorphism in the swordtail *Xiphophorus nigrensis*: testing the prediction of equal fitness', *American Naturalist*, January 1992, 139(1): 21–31.

Stephen M. Shuster, Michael J. Wade, 'Equal mating success among male reproductive strategies in a marine isopod', *Nature*, 18 April 1991, 350: 608–10.

Anna Runemark, Erik I. Svensson, 'Sexual selection as a promoter of population divergence in male phenotypic characters: a study on mainland and islet lizard populations', *Biological Journal of the Linnean Society*, 2012, 106: 374–89.

B. Sinervo, C. M. Lively, 'The rock-paper-scissors game and the evolution of alternative male strategies', *Nature*, 21 March 1996, 380: 240–3.

Thomas P. Gosden, Erik I. Svensson, 'Density-dependent male mating harassment, female resistance, and male mimicry', *American Naturalist*, June 2009, 173(6): 709–21.

Sven Steiner, Johannes L. M. Steidle, Joachim Ruther, 'Female sex pheromone in immature insect males – a case of pre-emergence chemical mimicry?' *Behavioral Ecology and Sociobiology*, 2005, 58: 111–20.

Michael Hrabar, Adela Danci, Paul W. Schaefer, Gerhard Gries, 'In the nick of time: males of the parasitoid wasp *Pimpla disparis* respond to semiochemicals from emerging mates', *Journal of Chemical Ecology*, 2012, 38: 253–61.

Joop Jukema, Theunis Piersma, 'Permanent female mimics in a lekking shorebird', *Biology Letters*, 2006, 2: 161–4.

R. Shine, T. Langkilde, R. T. Mason, 'Facultative pheromonal mimicry in snakes: "she-males" attract courtship only when it is useful', *Behavioral Ecology and Sociobiology*, 2012, 66: 691–5.

STIs

Cristina Virto, Carlos A. Zarate, Miguel Lopez-Ferber, Rosa Murillo, Primitivo Caballero, Trevor Williams, 'Gender-mediated differences in vertical transmission of a nucleopolyhedrovirus', *PLOS ONE*, August 2013, 8(8): e70932.

Jonathan J. Ryder, Mary-Jo Hoare, Daria Pastok, Michael Bottery, Michael Boots, Andrew Fenton, David Atkinson, Robert J. Knell, Gregory D. D. Hurst, 'Disease epidemiology in arthropods is altered by the presence of nonprotective symbionts', *American Naturalist*, March 2014, 183(3): E89–104.

Owen D. Seeman, Helen F. Nahrung, 'Female biased parasitism and the importance of host generation overlap in a sexually transmitted parasite of beetles', *Journal of Parasitology*, 2004, 90(1): 114–18.

Candice M. Mitchell, Susan Hutton, Garry S. A. Myers, Robert Brunham, Peter Timms, '*Chlamydia pneumoniae* is genetically diverse in animals and appears to have crossed the host barrier to humans on (at least) two occasions', *PLoS Pathogens*, May 2010, 6(5): e1000903.

Niels A. G. Kerstes, Camillo Bérénos, Oliver Y. Martin, 'Coevolving parasites and population size shape the evolution of mating behaviour', *BMC Evolutionary Biology*, 2013, 13: 29.

David P. Cowan, 'Life history and male dimorphism in the mite *Kennethiella trisetosa* (Acarina: Winterschmidtiidae), and its symbiotic relationship with the wasp *Ancistrocerus antelope* (Hymenoptera: Eumenidae)', *Annals of the Entomological Society of America*, 1984, 77(6): 725–32.

Imroze Khan, Nagaraj Guru Prasad, 'Male *Drosophila melanogaster* show adaptive mating bias in response to female infection status', *Journal of Insect Physiology*, 2013, 59: 1017–23.

Clare C. Rittschof, Swetapadma Pattanaik, Laura Johnson, Luis F. Matos, Jérémie Brusini, Marta L. Wayne, 'Sigma virus and male reproductive success in Drosophila melanogaster', *Behavioral Ecology and Sociobiology*, 2013, 67: 529–40.

K. Wojczulanis-Jakubas, M. Dynowska, D. Jakubas, 'Fungi prevalence in breeding pairs of a monogamous seabird – little auk, *Alle alle*', *Ethology, Ecology & Evolution*, 2011, 23(3): 240–7.

Adam Polkinghorne, Jon Hanger, Peter Timms, 'Recent advances in understanding the biology, epidemiology and control of chlamydial infections in koalas', *Veterinary Microbiology*, 2013, 165: 214–23.

Michael Jackson, Neil White, Phil Giffard, Peter Timms, 'Epizootiology of Chlamydia infections in two free-range koala populations', *Veterinary Microbiology*, 1999, 65: 255–64.

Joel White, Murielle Richard, Manuel Massot, Sandrine Meylan, 'Cloacal bacterial diversity increases with multiple mates: evidence of sexual transmission in female common lizards', *PLOS ONE*, July 2011, 6(7): e22339.

Melissah Rowe, Gabor Arpad Czirjak, Kevin J. McGraw, Mathieu Giraudeau, 'Sexual ornamentation reflects antibacterial activity of ejaculates in mallards', *Biology Letters*, 2011, 7: 740–2.

Patrick Abbot, Larry M. Dill, 'Sexually transmitted parasites and sexual selection in the milkweed leaf beetle, *Labidomera clivicollis*', *Oikos*, 2001, 92: 91–100.

Anders Pape Møller, 'A fungus infecting domestic flies manipulates sexual behaviour of its host', *Behavioral Ecology and Sociobiology*, 1993, 33: 403–7.

Deanna M. Soper, Lynda F. Delph, Curt M. Lively, 'Multiple paternity in the freshwater snail, *Potamopyrgus antipodarum*', *Ecology & Evolution*, 2012, 2(12): 3179–85.

Jesus Martinez-Padilla, Pablo Vergara, Francois Mougeot, Stephen M. Redpath, 'Parasitized mates increase infection risk for partners', *American Naturalist*, June 2012, 179(6): 811–20.

D. M. Soper, K. C. King, D. Vergara, C. M. Lively, 'Exposure to parasites increases promiscuity in a freshwater snail', *Biology Letters*, 2014, 10: 20131091.

Jonathan James Ryder, Daria Pastok, Mary-Jo Hoare, Michael J. Bottery, Michael Boots, Robert K. Knell, David Atkinson, Gregory D. D. Hurst, 'Spatial variation in food supply, mating behavior, and sexually transmitted disease epidemics', *Behavioral Ecology*, 2012, 24(3): 723–9.

Ann B. Lockhart, Peter H. Thrall, Janis Antonovics, 'Sexually transmitted diseases in animals: ecological and evolutionary implications', *Biological Reviews*, 1996, 71: 415–71.

Robert J. Knell, K. Mary Webberley, 'Sexually transmitted diseases of insects: distribution, evolution, ecology and host behaviour', *Biological Reviews*, 2004, 79: 557–81.

Richard H. Lambertsen, Barbara A. Kohn, John P. Sundberg, Claus D. Buergelt, 'Genital papillomatosis in sperm whale bulls', *Journal of Wildlife Diseases*, 1987, 23(3): 361–7.

Joel White, Pascal Mirleau, Etienne Danchin, Herve Mulard, Scott A. Hatch, Philipp Heeb, Richard H. Wagner, 'Sexually transmitted bacteria affect female cloacal assemblages in a wild bird', *Ecology Letters*, 2010, 13: 1515–24.

SECTION THREE: THE AFTERMATH

Plastic parents

Justin Charles Touchon, Karen Michelle Warkentin, 'Reproductive mode plasticity: Aquatic and terrestrial oviposition in a treefrog', *PNAS*, 27 May 2008, 105(21): 7495–9.

Armin Philipp Moczek, 'Facultative paternal investment in the polyphonic beetle *Onthophagus taurus*: the role of male morphology and social context', *Behavioral Ecology*, 1999, 10(6): 641–7.

J. J. Fontaine, T. E. Martin, 'Parent birds assess nest predation risk and adjust their reproductive strategies', *Ecology Letters*, 2006, 9: 428–34.

Shigeki Kishi, Takayoshi Nishida, 'Adjustment of Parental Investment in the Dung Beetle *Onthophagus atripennis* (Col., Scarabaeidae)', *Ethology*, 2006, 112: 1239–45.

Marcio Martins, Jose P. Pombal, Jr, Celio F. B. Haddad, 'Escalated aggressive behaviour and facultative parental care in the nest building gladiator frog, *Hyla faber*', *Amphibia-Reptilia*, 1998, 19: 65–73.

Wen-San Huang, Si-Min Lin, Sylvain Dubey, David A. Pike, 'Predation drives interpopulation differences in parental care expression', *Journal of Animal Ecology*, 2013, 82: 429–37.

B. Heulin, Y. Surget-Groba, A. Guiller, C. P. Guillaume, J. Deunff, 'Comparisons of mitochondrial DNA (mtDNA) sequences (16S rRNA gene) between oviparous and viviparous strains of *Lacerta vivipara*: a preliminary study', *Molecular Ecology*, 1999, 8: 1627–31.

J. Bleu, B. Heulin, C. Haussy, S. Meylan, M. Massot, 'Experimental evidence of early costs of reproduction in conspecific viviparous and oviparous lizards', *Journal of Evolutionary Biology*, 2012, 25: 1264–74.

Cameron K. Ghalambor, Susana I. Peluc, Thomas E. Martin, 'Plasticity of parental care under the risk of predation: how much should parents reduce care?', *Biology Letters*, 2013, 9: 20130154.

Nick J. Royle, Andrew F. Russell, Alastair Wilson, 'The evolution of flexible parenting', *Science*, 2014, 345: 776.

David Buckley, Marina Alcobendas, Mario Garcia-Paris, Marvalee H. Wake, 'Heterochrony, cannibalism, and the evolution of viviparity in *Salamandra salamandra*', *Evolution & Development*, 2007, 9(1): 105–15.

Liana Y. Zanette, Aija F. White, Marek C. Allen, Michael Clinchy, 'Perceived predation risk reduces the number of offspring songbirds produce per year', *Science*, 2011, 334: 1398.

Go ask your dad

David J. Hosken, Thomas H. Kunz, 'But is it male lactation or not?', *Trends in Ecology and Evolution*, 24(7), 355.

Daniel N. Racey, Malcolm Peaker, Paul A. Racey, 'Galactorrhoea is not lactation', *Trends in Ecology and Evolution*, 24(7), 354–5.

I. Ahnesjo, J. F. Craig, 'The biology of Syngnathidae: pipefishes, seadragons and seahorses', *Journal of Fish Biology*, 2011, 78: 1597–1602.

Olivia Roth, Verena Klein, Anne Beemelmanns, Jorn P. Scharsack, Thorsten B. H. Reusch, 'Male pregnancy and biparental immune priming', *American Naturalist*, December 2012, 180(6): 802–14.

Kimberly A. Paczolt, Adam G. Jones, 'Post-copulatory sexual selection and sexual conflict in the evolution of male pregnancy', *Nature*, 18 March 2010, 464: 401–4.

Lisa Jacquin, Lydie Blottiere, Claudy Haussy, Samuel Perret, Julien Gasparini, 'Prenatal and postnatal parental effects on immunity and growth in "lactating" pigeons', *Functional Ecology*, 2012, 26: 866–75.

Maria A. van Noordwijk, Erik P. Willems, Sri Suci Utami Atmoko, Christopher W. Kuzawa, Carel P. van Schaik, 'Multi-year lactation and its consequences in Bornean orangutans (*Pongo pygmaeus wurmbii*)', *Behavioral Ecology and Sociobiology*, 2013, 67: 805–14.

Thomas H. Kunz, David J. Hosken, 'Male lactation: why, why not and is it care?', *Trends in Ecology and Evolution*, 24(2): 80–5.

J. M. Stringer, A. J. Pask, G. Shaw, M. B, Renfree, 'Post-natal imprinting: evidence from marsupials', *Heredity*, 2014, 113: 145–55.

M. Cristina Keightley, Bob B. M. Wong, Graham J. Lieschke, 'Immune priming: mothering males modulate immunity', *Current Biology*, 23(2): R76–8.

C. Kvarnemo, K. B. Mobley, C. Partridge, A. G. Jones, I. Ahnesjo, 'Evidence of paternal nutrient provisioning to embryos in broad-nosed pipefish *Syngnathus typhle*', *Journal of Fish Biology*, 2011, 78: 1725–37.

Darryl T. Gwynne, Kevin A. Judge, Clint D. Kelly, 'Evidence for male allocation in pipefish?', *Nature*, 26 August 2010, 466: E11.

Kimberley A. Paczolt, Adam G. Jones, 'Paczolt & Jones reply', *Nature*, 26 August 2010, 466: E12.

Emily Rose, Kimberley A. Paczolt, Adam G. Jones, 'The contributions of premating and postmating selection episodes to total selection in sex-role-reversed gulf pipefish', *American Naturalist*, September 2013, 182(3): 410–20.

Anders Berglund, 'Pregnant fathers in charge', *Nature*, 18 March 2010, 464: 364–5.

G. Rosenqvist, A. Berglund, 'Sexual signals and mating patterns in Syngnathidae', *Journal of Fish Biology*, 2011, 78: 1647–61.

Extreme lactation

Louise Barrett, Jo Halliday, S. Peter Henzi, 'The ecology of motherhood: the structuring of lactation costs by chacma baboons', *Journal of Animal Ecology*, 2006, 75: 875–86.

Marilyn B. Renfree, 'Diapause, pregnancy, and parturition in Australian marsupials', *Journal of Experimental Zoology*, 1993, 266: 450–62.

J. M. Stringer, A. J. Pask, G. Shaw, M. B. Renfree, 'Post-natal imprinting: evidence from marsupials', *Heredity*, 2014, 113: 145–55.

Maria A. van Noordwijk, Erik P. Willems, Sri Suci Utami Atmoko, Christopher W. Kuzawa, Carel P. van Schaik, 'Multi-year lactation and its consequences in Bornean orangutans (*Pongo pygmaeus wurmbii*)', *Behavioral Ecology and Sociobiology*, 2013, 67: 805–14.

Sanjana Kuruppath, Swathi Bisana, Julie A. Sharp, Christophe Lefevre, Satish Kumar, Kevin R. Nicholas, 'Monotremes and marsupials: comparative models to better understand the function of milk', *Journal of Biosciences*, 2012, 37(4): 581–8.

Melanie J. Edwards, Janine E. Deakin, 'The marsupial pouch: implications for reproductive success and mammalian evolution', *Australian Journal of Zoology*, 2013, 61: 41–7.

Elie Khalil, Matthew R. Digby, Paul O'Donnell, Kevin R. Nicholas, 'Changes in milk protein composition during acute involution at different phases of tammar wallaby (*Macropus eugenii*) lactation', *Comparative Biochemistry and Physiology*, 2008, Part B 151: 64–9.

Uriel Gelin, Michelle E. Wilson, Graeme Coulson, Marco Festa-Bianchet, 'Experimental manipulation of female reproduction demonstrates its fitness costs in kangaroos', *Journal of Animal Ecology*, 2015, 84: 239–48.

Nanette Yvette Schneider, 'The development of the olfactory organs in newly hatched monotremes and neonate marsupials', *Journal of Anatomy*, 2011, 219: 229–42.

David Brawand, Walter Wahli, Henrik Kaessmann, 'Loss of egg yolk genes in mammals and the origin of lactation and placentation', *PLOS Biology*, March 2008, 6(3): e63.

Coralie M. Reich, John P. Y. Arnould, 'Evolution of Pinnipedia lactation strategies: a potential role for α-lactalbumin?' *Biology Letters*, 2007, 3: 546–9.

M. J. Edwards, L. A. Hinds, E. M. Deane, J. E. Deakin, 'A review of complementary mechanisms which protect the developing marsupial pouch young', *Developmental and Comparative Immunology*, 2012, 37: 213–20.

Karen Fey, Fritz Trillmich, 'Sibling competition in guinea pigs (*Cavia aperea f. porcellus*): scrambling for mother's teats is stressful', *Behavioral Ecology and Sociobiology*, 2008, 62: 321–9.

Robyn Hudson, Hans Distel, 'Fighting by kittens and piglets during suckling: what does it mean?', *Ethology*, 2013, 119: 353–9.

Nanette Y. Schneider, Terrence P. Fletcher, Geoff Shaw, Marilyn B. Renfree, 'The effect of pregnant and oestrous females on male testosterone and behaviour in the tammar wallaby', *Hormones and Behavior*, 2010, 58: 378–84.

Nanette Y. Schneider, Terrence P. Fletcher, Geoff Shaw, Marilyn B. Renfree, 'The olfactory system of the tammar wallaby is developed at birth and directs the neonate to its mother's pouch odours', *Reproduction*, 2009, 138: 849–57.

Set yourself up for success

Suvi Ruuskanen, Blandine Doligez, Natalia Pitala, Lars Gustafsson, Toni Laaksonen, 'Long-term fitness consequences of high yolk androgen levels: sons pay the costs', *Functional Ecology*, 2012, 26: 884–94.

Jan Kristofi k, Alzbeta Darolova, Juraj Majtan, Monika Okuliarova, Michal Zeman, Herbert Hoi, 'Do females invest more into eggs when males sing more attractively? Postmating sexual selection strategies in a monogamous reed passerine', *Ecology & Evolution*, 2014, 4(8): 1328–39.

Fabrice Dentressangle, Lourdes Boeck, Roxana Torres, 'Maternal investment in eggs is affected by male feet colour and breeding conditions in the blue-footed booby, Sula nebouxii', *Behavioral Ecology and Sociobiology*, 2008, 62: 1899–1908.

V. Garcia-Fernandez, T. I. Draganoiu, D. Ung, A. Lacroix, G. Malacarne, G. Leboucher, 'Female canaries invest more in response to an exaggerated male trait', *Animal Behaviour*, 2013, 85: 679–84.

Carlos Alonso-Alvarez, Lorenzo Pérez-Rodríguez, María Ester Ferrero, Esther García de-Blas, Fabián Casas, Francois Mougeot, 'Adjustment of female reproductive investment according to male carotenoid-based ornamentation in a gallinaceous bird', *Behavioral Ecology and Sociobiology*, 2012, 66: 731–42.

Michael Coslovsky, Ton Groothuis, Bonnie de Vries, Heinz Richner, 'Maternal steroids in egg yolk as a pathway to translate predation risk to offspring: Experiments with great tits', *General and Comparative Endocrinology*, 2012, 176: 211–14.

Zachary R. Stahlschmidt, Richard Shine, Dale F. DeNardo, 'The consequences of alternative parental care tactics in free-ranging pythons in tropical Australia', *Functional Ecology*, 2012, 26: 812–21.

Rene E. van Dijk, Corine M. Eising, Richard M. Merrill, Filiz Karadas, Ben Hatchwell, Claire N. Spottiswoode, 'Maternal effects in the highly communal sociable weaver may exacerbate brood reduction and prepare offspring for a competitive social environment', *Oecologia*, 2013, 171: 379–89.

Helga Gwinnera, Elizabeth Yohannesa, Hubert Schwab, 'Nest composition and yolk hormones: do female European starlings adjust yolk androgens to nest quality?', *Avian Biology Research*, 2013, 6(4): 307–12.

Marco Grenna, Lorena Avidano, Giorgio Malacarne, Gerard Leboucher, Marco Cucco, 'Influence of male dominance on egg testosterone and antibacterial substances in the egg of grey partridges', *Ethology*, 2013, 120: 149–58.

Suvi Ruuskanen, Toni Laaksonen, 'Sex-specific effects of yolk androgens on begging behavior and digestion in pied flycatchers *Ficedula hypoleuca*', *Journal of Avian Biology*, 2013, 44: 331–38.

Gergely Hegyi, Marton Herenyi, Eszter Szollosi, Balazs Rosivall, Janos Torok, Ton G. G. Groothuis, 'Yolk androstenedione, but not testosterone, predicts offspring fate and reflects parental quality', *Behavioral Ecology*, 2011, 22: 29–38.

Susan C. Grana, Scott K. Sakaluk, Rachel M. Bowden, Melissa A. Doellman, Laura A. Vogel, Charles F. Thompson, 'Reproductive allocation in female house wrens is not influenced by experimentally altered male attractiveness', *Behavioral Ecology and Sociobiology*, 2012, 66: 1247–58.

Vladimír Remeš, 'Yolk androgens in great tit eggs are related to male attractiveness, breeding density and territory quality', *Behavioral Ecology and Sociobiology*, 2011, 65: 1257–66.

Take my kids. No really, TAKE THEM!

Marie E. Herberstein, Heather J. Baldwin, Anne C. Gaskett, 'Deception down under: is Australia a hot spot for deception?', *Behavioral Ecology*, 2014, 25(1): 12–16.

Ros Gloag, Alex Kacelnik, 'Host manipulation via begging call structure in the brood-parasitic shiny cowbird', *Animal Behaviour*, 2013, 86: 101–9.

Marcel Honza, Michal Sulc, Václav Jelínek, Milica Pozgayová, Petr Procházka, 'Brood parasites lay eggs matching the appearance of host clutches', *Proceedings of the Royal Society B*, 2014, 281: 20132665.

Simon Ducatez, 'Brood parasitism: a good strategy in our changing world?' *Proceedings of the Royal Society B*, 2014, 281: 20132404.

Vanina D. Fiorini, Ros Gloag, Alex Kacelnik, Juan C. Reboreda, 'Strategic egg destruction by brood-parasitic cowbirds?', *Animal Behaviour*, 2014, 93: 229–35.

Jay R. Stauffer, Jr, William F. Loftus, 'Brood Parasitism of a Bagrid Catfish (Bagrus *meridionalis*) by a Clariid Catfish (Bathyclarias *nyasensis*) in Lake Malaŵi, Africa', *Copeia*, 2010, 2010(1): 71–4.

Csaba Moskát, Márk E. Hauber, Zoltán Elek, Moniek Gommers, Miklós Bán, Frank Groenewoud, Tom S. L. Versluijs, Christiaan W. A. Hoetz, Jan Komdeur, 'Foreign egg retention by avian hosts in repeated brood parasitism: why do rejecters accept?', *Behavioral Ecology and Sociobiology*, 2014, 68: 403–13.

Maria Roldan, Manuel Soler, 'Parental-care parasitism: how do unrelated offspring attain acceptance by foster parents?', *Behavioral Ecology*, 2011, 22: 679–91.

Daniela Canestrari, Diana Bolopo, Ted C. J. Turlings, Gregory Röder, José M. Marcos, Vittorio Baglione, 'From parasitism to mutualism: unexpected interactions between a cuckoo and its host', *Science*, 2014, 343: 1350.

Hannu Pöysä, Antti Paasivaara, Kari Lindblom, Jarkko Rutila, Jorma Sorjonen, 'Co-parasites preferentially lay with kin and in safe neighbourhoods: experimental evidence from goldeneye ducks', *Animal Behaviour*, 2014, 91: 111–18.

Michael D. Sorenson, Kristina M. Sefc, Robert B. Payne, 'Speciation by host switch in brood parasitic indigobirds', *Nature*, 21 August 2003, 424: 928–31.

Manuel Soler, Liesbeth de Neve, Gianluca Roncalli, Elena Macías-Sánchez, Juan Diego Ibáñez-Álamo, Tomás Pérez-Contreras, 'Great spotted cuckoo fledglings are disadvantaged by magpie host parents when reared together with magpie nestlings', *Behavioral Ecology and Sociobiology*, 2014, 68: 333–42.

Ros Gloag, Vanina D. Fiorini, Juan C. Reboreda, Alex Kacelnik, 'The wages of violence: mobbing by mockingbirds as a frontline defence against brood-parasitic cowbirds', *Animal Behaviour*, 2013, 86: 1023–9.

M. P. Haesler, C. M. Lindeyer, M. Taborsky, 'Reproductive parasitism: male and female responses to conspecific and heterospecific intrusions at spawning in a mouth-brooding cichlid *Ophthalmotilapia ventralis*', *Journal of Fish Biology*, 2009, 75: 1845–56.

Tetsu Sato, 'A brood parasitic catfish of mouthbrooding cichlid fishes in Lake Tanganyika', *Nature*, 4 September 1986, 323: 58–9.

Gregory Röder, Daniela Canestrari, Diana Bolopo, José M. Marcos, Neil Villard, Vittorio Baglione, Ted C. J. Turlings, 'Chicks of the great spotted cuckoo may turn brood parasitism into mutualism by producing a foul-smelling secretion that repels predators', *Journal of Chemical Ecology*, 2014, 40: 320–4.

Everton Tizo-Pedrosoa, Kleber Del-Claro, 'Social parasitism: emergence of the cuckoo strategy between pseudoscorpions', *Behavioral Ecology*, 2014, 25(2): 335–43.

Ros Gloag, Vanina D. Fiorini, Juan Carlos Reboreda, Alex Kacelnik, 'Shiny cowbirds share foster mothers but not true mothers in multiply parasitized mockingbird nests', *Behavioral Ecology and Sociobiology*, 2014, 68: 681–9.

You should not kill a child

S. M. J. G. Steyaert, C. Reusch, S. Brunberg, J. E. Swenson, K. Hackländer, A. Zedrosser, 'Infanticide as a male reproductive strategy has a nutritive risk effect in brown bears', *Biology Letters*, 2013, 9: 20130624.

J. A. Oldekop, P. T. Smiseth, H. D. Piggins, A. J. Moore, 'Adaptive switch from infanticide to parental care: how do beetles time their behaviour?', *Journal Evolutionary Biology*, 2007, 20(5): 1998–2004.

Julie A. Teichroeb, Pascale Sicotte, 'Cost-free vigilance during feeding in folivorous primates? Examining the effect of predation risk, scramble competition, and infanticide threat on vigilance in ursine colobus monkeys (*Colobus vellerosus*)', *Behavioral Ecology and Sociobiology*, 2012, 66: 453–66.

Reinmar Hager, Rufus A. Johnstone, Infanticide and control of reproduction in cooperative and communal breeders', *Animal Behaviour*, 2004, 67: 941–9.

Meeghan E. Gray, Elissa Z. Cameron, Mary M. Peacock, David S. Thain, Veronica S. Kirchoff, 'Are low infidelity rates in feral horses due to infanticide?', *Behavioral Ecology and Sociobiology*, 2012, 66: 529–37.

Jep Agrell, Jerry O. Wolff, Hannu Ylonen, 'Counter-strategies to infanticide in mammals: costs and consequences', *Oikos*, 1998, 83(3): 507–17.

Carola Borries, Tommaso Savini, Andreas Koenig, 'Social monogamy and the threat of infanticide in larger mammals, *Behavioral Ecology and Sociobiology*, 2011, 65: 685–93.

Qing Zhao, Carola Borries, Wenshi Pan, 'Male takeover, infanticide, and female countertactics in white-headed leaf monkeys (*Trachypithecus leucocephalus*)', *Behavioral Ecology and Sociobiology*, 2011, 65: 1535–47.

Andrew J. Young, Tim Clutton-Brock, 'Infanticide by subordinates influences reproductive sharing in cooperatively breeding meerkats,' *Biology Letters*, 2006, 2: 385–7.

Arne Janssen, Farid Faraji, Tessa van der Hammen, Sara Magalhaes, Maurice W. Sabelis, 'Interspecific infanticide deters predators', *Ecology Letters*, 2002, 5: 490–4.

Alfréd Trnkal, Pavol Prokop, Péter Batáry, 'Infanticide or interference: Does the great reed warbler selectively destroy eggs?', *Annales Zoologici Fennici*, 2010, 47: 272–7.

Mirjam Knornschild, Katja Ueberschaer, Maria Helbig, Elisabeth K. V. Kalko, 'Sexually selected infanticide in a polygynous bat', *PLOS ONE*, September 2011, 6(9): e25001.

A. David M. Latham, Stan Boutin, 'Wolf, *Canis lupus*, pup mortality: interspecific predation or non-parental infanticide?', *Canadian Field-Naturalist*, 2011, 25(2): 158–61.

Child abuse

Jacquelyn K. Grace, Karen Dean, Mary Ann Ottinger, David J. Anderson, 'Hormonal effects of maltreatment in Nazca booby nestlings: Implications for the "cycle of violence"', *Hormones and Behavior*, 2011, 60: 78–85.

Kate Ashbrook, Sarah Wanless, Mike P. Harris and Keith C. Hamer, 'Hitting the buffers: conspecific aggression undermines benefits of colonial breeding under adverse conditions', *Biology Letters*, 2008, 4: 630–33.

Christel C. Bastida, Frank Puga, Francisco Gonzalez-Lima, Kimberly J. Jennings, Joel C. Wommack, Yvon Delville, 'Chronic social stress in puberty alters appetitive male sexual behavior and neural metabolic activity', *Hormones and Behavior*, 2014, 66, 220–7.

David J. Anderson, Elaine T. Porter, Elise D. Ferree, 'Non-breeding Nazca boobies (*Sula granti*) show social and sexual interest in chicks: behavioural and ecological aspects', *Behaviour*, 2004, 141(8): 959–77.

Christopher M. Somers, Victoria A. Kjoss, R. Mark Brigham, 'American white pelicans force copulations with nestlings', *Wilson Journal of Ornithology*, 2007, 119(2): 279–83.

Martina S. Müller, Elaine T. Porter, Jacquelyn K. Grace, Jill A. Awkerman, Kevin T. Birchler, Alex R. Gunderson, Eric G. Schneider, Mark A. Westbrock, David J. Anderson, 'Maltreated Nestlings Exhibit Correlated Maltreatment as Adults: Evidence of a "Cycle of Violence" in Nazca Boobies (*Sula granti*)', *The Auk*, 2011, 128(4): 615–19.

Tim Burton, Neil B. Metcalfe, 'Can environmental conditions experienced in early life influence future generations?', *Proceedings of the Royal Society B*, 2014, 281: 20140311.

Yukiko Inoue, Ken Yoda, Hidenori Fujii, Hirofumi Kuroki, Yasuaki Niizuma, 'Nest intrusion and infanticidal attack on nestlings in great cormorants Phalacrocorax carbo: why do adults attack conspecific chicks?', *Journal of Ethology*, 2010, 28: 221–30.

Siblicide

Martina Muller, Ton G. G. Groothuis, 'Within-clutch variation in yolk testosterone as an adaptive maternal effect to modulate avian sibling competition: evidence from a comparative study', *American Naturalist*, January 2013, 181(1): 125–36.

Antoni Margalida, Diego García, Rafael Heredia, Joan Bertran, 'Video-monitoring helps to optimize the rescue of second-hatched chicks in the endangered Bearded Vulture *Gypaetus barbatus*', *Bird Conservation International*, 2010, 20: 55–61.

Hugh Drummond, Cristina Rodríguez, 'No reduction in aggression after loss of a broodmate: a test of the brood size hypothesis', *Behavioral Ecology and Sociobiology*, 2009, 63: 321–7.

Ola M. Fincke, 'Excess offspring as a maternal strategy: constraints in the shared nursery of a giant damselfly', *Behavioral Ecology*, 2011, 22: 543–51.

Ralph Dobler, Mathias Kölliker, 'Influence of weight asymmetry and kinship on siblicidal and cannibalistic behaviour in earwigs', *Animal Behaviour*, 2011, 667–72.

Ralph Dobler, Mathias Kölliker, 'Kin-selected siblicide and cannibalism in the European earwig', *Behavioral Ecology*, 2010, 21: 257–63.

Fritz Trillmich, Jochen B. W. Wolf, 'Parent–offspring and sibling conflict in Galápagos fur seals and sea lions', *Behavioral Ecology and Sociobiology*, 2008, 62: 363–75.

Heribert Hofer, Marion L. East, 'Siblicide in Serengeti spotted hyenas: a long-term study of maternal input and cub survival', *Behavioral Ecology and Sociobiology*, 2008, 62: 341–51.

Martina S. Müller, Børge Moe, Ton G. G. Groothuis, 'Testosterone increases siblicidal aggression in black-legged kittiwake chicks (*Rissa tridactyla*)', *Behavioral Ecology and Sociobiology*, 2014, 68: 223–32.

Sarah Benhaiem, Heribert Hofer, Martin Dehnhard, Janine Helms, Marion L. East, 'Sibling competition and hunger increase allostatic load in spotted hyaenas', *Biology Letters*, 2013, 9: 20130040.

Ning Wang, Rebecca T. Kimball, 'Nestmate killing by obligate brood parasitic chicks: is this linked to obligate siblicidal behavior?', *Journal of Ornithology*, 2012, 153: 825–31.

Apostolos Kapranas, Ian C. W. Hardy, Joseph G. Morse, Robert F. Luck, 'Parasitoid developmental mortality in the field: patterns, causes and consequences for sex ratio and virginity', *Journal of Animal Ecology*, 2011, 80: 192–203.

David W. Pfennig, Paul W. Sherman, 'Kin recognition and cannibalism in polyphonic salamanders', *Behavioral Ecology*, 1994, 5: 225–32.

Thomas Merkling, Lena Agdere, Elise Albert, Romain Durieux, Scott A. Hatch, Etienne Danchin, Pierrick Blanchard, 'Is natural hatching asynchrony optimal? An experimental investigation of sibling competition patterns in a facultatively siblicidal seabird', *Behavioral Ecology and Sociobiology*, 2014, 68: 309–19.

Murray P. Fea, Margaret C. Stanley, Gregory I. Holwell, 'Cannibalistic siblicide in praying mantis nymphs (*Miomantis caffra*)', *Journal of Ethology*, 2014, 32: 43–51.

Sarah Benhaiem, Heribert Hofer, Stephanie Kramer-Schadt, Edgar Brunner, Marion L. East, 'Sibling rivalry: training effects, emergence of dominance and incomplete control', *Proceedings of the Royal Society B*, 2012, 279: 3727–35.

Maria Modanua, Lucy Dong Xuan Li, Hosay Said, Nizanthan Rathitharan, Maydianne C. B. Andrade, 'Sibling cannibalism in a web-building spider: Effects of density and shared environment', *Behavioural Processes*, 2014, 106: 12–16.

David W. Pfennig, Hudson K. Reeve, Paul W. Sherman, 'Kin recognition and cannibalism in spadefoot toad tadpoles', *Animal Behaviour*, 1993, 46: 87–94.

Clutch piracy

Laslo Demeter, Zoltan Benko, 'Male *Rana temporaria* in amplexus with a clutch', *North-Western Journal of Zoology*, 2007, 3(2): 105–8.

Jason L. Brown, Victor Morales, Kyle Summers, 'Tactical reproductive parasitism via larval cannibalism in Peruvian poison frogs', *Biology Letters*, 2009, 5: 148–51.

David R. Vieites, Sandra Nieto-Roman, Marta Barluenga, Antonio Palanca, Miguel Vences, Axel Meyer, 'Post-mating clutch piracy in an amphibian', *Nature*, 16 September 2004, 431: 305–8.

Arrested development

Andrea Friebe, Alina L. Evans, Jon M. Arnemo, Stephane Blanc, Sven Brunberg, Gunther Fleissner, Jon E. Swenson, Andreas Zedrosser, 'Factors Affecting Date of Implantation, Parturition, and Den Entry Estimated from Activity and Body Temperature in Free-Ranging Brown Bears', *PLOS ONE*, July 2014, 9(7): e101410.

Helena Larsdotter-Mellström, Rushana Murtazina, Anna-Karin Borg-Karlson, Christer Wiklund, 'Timing of male sex pheromone biosynthesis in a butterfly – different dynamics under direct or diapause development', *Journal of Chemical Ecology*, 2012, 38: 584–91.

Irma Varela-Lasheras, Tom J. M. Van Dooren, 'Desiccation plasticity in the embryonic life histories of non-annual rivulid species', *EvoDevo*, 2014, 5: 16.

Jane C. Fenelon, Arnab Banerjee, Bruce D. Murphy, 'Embryonic diapause: development on hold', *International Journal of Developmental Biology*, 2014, 58: 163–74.

Geoffrey Fryer, 'Diapause, a potent force in the evolution of freshwater crustaceans', *Hydrobiologia*, 1996, 320: 1–14.

Nadiya V. Evseeva, 'Diapause of copepods as an element for stabilizing the parasite system of some fish helminths', *Hydrobiologia*, 1996, 320: 229–33.

M. Slusarczyk, A. Ochocka, D. Cichocka, 'The prevalence of diapause response to risk of size-selective predation in small- and large-bodied prey species', *Aquatic Ecology*, 2012, 45: 1–8.

Philipp Zerbe, Marcus Clauss, Daryl Codron, Laurie Bingaman Lackey, Eberhard Rensch, Jurgen W. Streich, Jean-Michel Hatt,

Dennis W. H. Muller, 'Reproductive seasonality in captive wild ruminants: implications for biogeographical adaptation, photoperiodic control, and life history', *Biological Reviews*, 2012, 87: 965–90.

D. Fordham, A. Georges, B. Corey, 'Compensation for inundation-induced embryonic diapause in a freshwater turtle: achieving predictability in the face of environmental stochasticity', *Functional Ecology*, 2006, 20: 670–77.

B. D. Murphy, 'Embryonic Diapause: Advances in Understanding the Enigma of Seasonal Delayed Implantation', *Reproduction in Domestic Animals*, 2012, 47(Suppl. 6): 121–4.

Grazyna E. Ptak, Jacek A. Modlinski, Pasqualino Loi, 'Emryonic diapause in humans: time to consider?', *Reproductive Biology and Endocrinology*, 2013, 11: 92.

C. A. Moraiti, C. T. Nakas, N. T. Papadopoulos, 'Diapause termination of *Rhagoletis cerasi* pupae is regulated by local adaptation and phenotypic plasticity: escape in time through bet-hedging strategies', *Journal of Evolutionary Biology*, 2014, 27: 45–54.

Grazyna E. Ptak, Emanuela Tacconi, Marta Czernik, Paola Toschi, Jacek A. Modlinski, Pasqualino Loi, 'Embryonic diapause is conserved across mammals', *PLOS ONE*, March 2012, 7(3): e33027.

Panu Välimäki, Sami M. Kivelä, Maarit I. Mäenpää, 'Temperature- and density-dependence of diapause induction and its life history correlates in the geometrid moth *Chiasmia clathrata* (Lepidoptera: Geometridae)', *Evolutionary Ecology*, 2013, 27: 1217–33.

Elham Sheikh-Jabbari, Matthew D. Hall, Frida Ben-Ami, Dieter Ebert, The expression of virulence for a mixed-mode transmitted parasite in a diapausing host', *Parasitology*, 2014, 141: 1097–1107.

A. W. Park, J. Vandekerkhove, Y. Michalakis, 'Sex in an uncertain world: environmental stochasticity helps restore competitive balance between sexually and asexually reproducing populations', *Journal of Evolutionary Biology*, 2014, 27: 1650–61.

W. Charles Kerfoot, Foad Yousef, Martin M. Hobmeier, Ryan P. Maki, S. Taylor Jarnagin, James H. Churchill, 'Temperature, recreational fishing and diapause egg connections: dispersal of spiny water fleas (*Bythotrephes longimanus*)', *Biological Invasions*, 2011, 13: 2513–31.

Jerry A. Powell, 'Synchronized, mass-emergences of a yucca moth, *Prodoxus Y-inversus* (Lepidoptera: Prodoxidae), after 16 and 17 years in diapause', *Oecologia*, 1989, 81: 490–3.

Tired old ladies and dirty old men

Daniel H. Nussey, Hannah Froy, Jean-Francois Lemaitre, Jean-Michel Gaillard, Steve N. Austad, 'Senescence in natural populations of animals: Widespread evidence and its implications for bio-gerontology', *Ageing Research Reviews*, 2013, 12: 214–25.

Deborah Pardo, Christophe Barbraud, Henri Weimerskirch, 'What shall I do now? State-dependent variations of life-history traits with aging in Wandering Albatrosses', *Ecology and Evolution*, 2014, 4(4): 474–87.

Pau Carazo, Pau Molina-Vila, Enrique Font, 'Male reproductive senescence as a potential source of sexual conflict in a beetle', *Behavioral Ecology*, 2011, 22: 192–8.

Roxana Torres, Alberto Velando, 'Male reproductive senescence: the price of immune- induced oxidative damage on sexual attractiveness in the blue-footed booby', *Journal of Animal Ecology*, 2007, 76: 1161–8.

Jean-François Lemaître, Jean-Michel Gaillard, Josephine M. Pemberton, Tim H. Clutton-Brock, Daniel H. Nussey, 'Early life expenditure in sexual competition is associated with increased reproductive senescence in male red deer', *Proceedings of the Royal Society B*, 2014, 281: 20140792.

A. D. Hayward, K. U. Mar, M. Lahdenpera, V. Lummaa, 'Early reproductive investment, senescence and lifetime reproductive success in female Asian elephants', *Journal of Evolutionary Biology*, 2014, 27: 772–83.

Deborah Pardo, Christophe Barbraud, Henri Weimerskirch, 'Females better face senescence in the wandering albatross', *Oecologia*, 2013, 173: 1283–94.

Gwenael Beauplet, Christophe Barbraud, Willy Dabin, Clothilde Kussener, Christophe Guinet, 'Age-specific survival and reproductive performances in fur seals: evidence of senescence and individual quality', *Oikos*, 2006, 112: 430–41.

Stuart P. Sharp, Tim H. Clutton-Brock, 'Reproductive senescence in a cooperatively breeding mammal', *Journal of Animal Ecology*, 2010, 79, 176–83.

Emily H. DuVal, 'Variation in annual and lifetime reproductive success of lance-tailed manakins: alpha experience mitigates effects of senescence on siring success', *Proceedings of the Royal Society B*, 2012, 279: 1551–9.

Christina M. Schmidt, Jennifer U.M. Jarvis, Nigel C. Bennett, 'The long-lived queen: reproduction and longevity in female eusocial Damaraland mole-rats (Fukomys *damarensis*)', *African Zoology*, 2013, 48(1): 193–6.

Philip Dammann, Radim Sumbera, Christina Maßmann, Andre Scherag, Hynek Burda, 'Extended longevity of reproductives appears to be common in *Fukomys* mole-rats (Rodentia, Bathyergidae)', *PLOS ONE*, April 2011, 6(4): e18757.

Christina M. Schmidt, Jonathan D. Blount, Nigel C. Bennett, 'Reproduction is associated with a tissue-dependent reduction of oxidative stress in eusocial female Damaraland mole-rats (Fukomys damarensis)', *PLOS ONE*, July 2014, 9(7): e103286.

Robert B. Weladji, Øystein Holand, Jean-Michel Gaillard, Nigel G. Yoccoz, Atle Mysterud, Mauri Nieminen, Nils C. Stenseth, 'Age-specific changes in different components of reproductive output in female reindeer: terminal allocation or senescence?', *Oecologia*, 2010, 162: 261–71.

Sin-Yeon Kim, Alberto Velando, Roxana Torres, Hugh Drummond, 'Effects of recruiting age on senescence, lifespan and lifetime reproductive success in a long-lived seabird', *Oecologia*, 2011, 166: 615–26.

Thomas E. Reed, Loeske E. B. Kruuk, Sarah Wanless, Morten Frederiksen, Emma J. A. Cunningham, Michael P. Harris, 'Reproductive senescence in a long-lived seabird: rates of decline in late-life performance are associated with varying costs of early reproduction', *American Naturalist*, February 2008, 171(2): E89–101.

Mirkka Lahdenperä, Khyne U. Mar, Virpi Lummaa, 'Reproductive cessation and post-reproductive lifespan in Asian elephants and pre-industrial humans', *Frontiers in Zoology*, 2014, 11: 54.

Josh R. Auld, John F. Henkel, 'Diet alters delayed selfing, inbreeding depression, and reproductive senescence in a freshwater snail', *Ecology and Evolution*, 2014, 4(14): 2968–2977.

Adam D. Hayward, Alastair J. Wilson, Jill G. Pilkington, Tim H. Clutton-Brock, Josephine M. Pemberton, Loeske E. B. Kruuk, 'Reproductive senescence in female Soay sheep: variation across traits and contributions of individual ageing and selective disappearance', *Functional Ecology*, 2013, 27: 184–95.

Sebastien Descamps, Stan Boutin, Dominique Berteaux, Jean-Michel Gaillard, 'Age-specific variation in survival, reproductive success and offspring quality in red squirrels: evidence of senescence', *Oikos*, 2008, 117: 1406–16.

Hannah Froy, Richard A. Phillips, Andrew G. Wood, Daniel H. Nussey Sue Lewis, 'Age-related variation in reproductive traits in the wandering albatross: evidence for terminal improvement following senescence', *Ecology Letters*, 2013, 16: 642–9.

Jelle J. Boonekamp, Martijn Salomons, Sandra Bouwhuis, Cor Dijkstra, Simon Verhulst, 'Reproductive effort accelerates actuarial senescence in wild birds: an experimental study', *Ecology Letters*, 2014, 17: 599–605.

Grandmamas

David Lusseau, Karsten Schneider, Oliver J. Boisseau, Patti Haase, Elisabeth Slooten, Steve M. Dawson, 'The bottlenose dolphin community of Doubtful Sound features a large proportion of long-lasting associations: Can geographic isolation explain this unique trait?', *Behavioral Ecology and Sociobiology*, 2003, 54: 396–405.

Daryl P. Shanley, Rebecca Sear, Ruth Mace, Thomas B. L. Kirkwood, 'Testing evolutionary theories of menopause', *Proceedings of the Royal Society B*, 2007, 274: 2943–9.

Rufus A. Johnstone, Michael A. Cant, 'The evolution of menopause in cetaceans and humans: the role of demography', *Proceedings of the Royal Society B*, 2010, 277: 3765–71.

Helen Perich Alvarez, 'Grandmother Hypothesis and Primate Life Histories', *American Journal of Physical Anthropology*, 2000, 113: 435–50.

M. Lahdenperä, A. F. Russell, V. Lummaa, 'Selection for long lifespan in men: benefits of grandfathering?', *Proceedings of the Royal Society B*, 2007, 274: 2437–44.

Molly Fox, Rebecca Sear, Jan Beise, Gillian Ragsdale, Eckart Voland, Leslie A. Knapp, 'Grandma plays favourites: X-chromosome relatedness and sex-specific childhood mortality', *Proceedings of the Royal Society B*, 2010, 277: 567–73.

Susan C. Alberts, Jeanne Altmann, Diane K. Brockman, Marina Cords, Linda M. Fedigan, Anne Pusey, Tara S. Stoinski, Karen B. Strier, William F. Morris, Anne M. Bronikowski, 'Reproductive aging patterns in primates reveal that humans are distinct', *PNAS*, 13 August 2013, 110(33): 13440–45.

Kristen Hawkes, James E. Coxworth, 'Grandmothers and the evolution of human longevity: a review of findings and future directions', *Evolutionary Anthropology*, 2013, 22: 294–302.

Daniel A. Levitis, Oskar Burger, Laurie Bingaman Lackey, 'The human post-fertile lifespan in comparative evolutionary context', *Evolutionary Anthropology*, 2013, 22: 66–79.

Bernard Crespi, 'The insectan apes', *Human Nature*, 2014, 25: 6.

Mary S. Macdonald Pavelka, Linda M. Fedigan, Sandra Zohar, 'Availability and adaptive value of reproductive and postreproductive Japanese macaque mothers and grandmothers', *Animal Behaviour*, 2002, 64: 407–14.

Emma A. Foster, Daniel W. Franks, Sonia Mazzi, Safi K. Darden, Ken C. Balcomb, John K. B. Ford, Darren P. Croft, 'Adaptive Prolonged Postreproductive Life Span in Killer Whales', *Science*, 14 September 2012, 337: 1313.

Peter S. Kim, James E. Coxworth, Kristen Hawkes, 'Increased longevity evolves from grandmothering', *Proceedings of the Royal Society B*, 2012, 279: 4880–4.

Eric J. Ward, Kim Parsons, Elizabeth E. Holmes, Ken C. Balcomb III, John K. B. Ford, 'The role of menopause and reproductive senescence in a long-lived social mammal', *Frontiers in Zoology*, 2009, 6: 4.